COGNITIV
DIALECTICAL
BEHAVIOR THERAPY

The Ultimate CBT and DBT Guide to
Interpersonal Effectiveness, Emotion
Regulation, Cognitive Dissonance, PTSD, Panic,
Worry, Anxiety, and Self-Compassion
With
David A. Gillihan PhD
Jason J. Shepherd M.D.
Jeffrey Satterfield PsyD

Seth Clark PhD
Matthew Wood M.D.
Marsha Bradberry PsyD
James Linehan PsyD

Table of Contents

Introduction

It is only now in our modern world of technology that we have just begun to scratch the surface of one of the greatest machines ever used. Our brains and the power of our thoughts is the one instrument that has been used for millennia to control the world. Nothing that we do in this world happens unless it happens first in our thoughts.

Throughout the centuries, it has been noted and referred to repeatedly in every culture of the world. It was Buddha who said, "All that we are arises with our thoughts." King Solomon was quoted as saying, "For as he thinks in his heart, so is he." Even the Qur'an referred to it in the scripture, "Surely Allah does not change the condition of a people until they change their own condition."

For countless years, we have understood this fundamental basis of human nature, and for centuries, it was addressed in many Eastern cultures through meditation, mindfulness, and other forms of spiritual connections. Still, it has only been in the past century that the Western civilization has recognized that the flow of our thoughts running through our minds is at the very core of human behavior, and as a result, has used this fundamental truth as the basis for a new form of psychotherapy called Cognitive Behavioral Therapy.

Research studies have shown that Cognitive Behavioral Therapy (CBT) is highly effective in treating a number of depression and anxiety disorders. It has been touted as even more effective than medication or other forms of psychotherapy. However, it is still quite challenging for the layperson to appreciate the value of this form of treatment until they have actually seen it put to use.

People who suffer from anxiety and depressive disorders often struggle to find a lasting solution to their problems. They read book after book filled with technical jargon that they find difficult to understand and end up feeling as if getting relief is hopeless; they are exhausted and confused. However, here in the following pages, we will attempt to break down the basic concept of CBT into simple terms that everyone can understand, thus arming you with the power to heal yourself and correct those intrusive thoughts that have permeated your life.

This book is designed to give you a way to replace that constant flow of thoughts that are inhibiting your ability to achieve the kind of behavior you want and replace them with something that will empower you instead. It speaks to a team effort that helps us to find what is broken in our minds and provide the tools to fix it through a logical and systematic approach. Together you will learn:

- What CBT really is
- Who it can help and how
- Some of the common causes for developing anxiety and depressive disorders
- How better to understand how the mind works and what triggers negative thoughts
- Practical steps to treat different types of disorders
- Different types of therapies and how they can help
- And so much more

Whether you're looking for CBT to help with a personal problem or you want to give guidance to someone you care about, you'll learn here how to be the best friend to yourself or those around you and to bring out the positive elements that exist in all of us. Can you benefit from CBT? Chances are good that you can. So as long as we are both here together, let's work this treatment out and get to the root of the problems you're facing. My guess is that if you're ready to learn something new about yourself and ready to apply it, you'll grow from this knowledge and be able to change that constant stream of thoughts that are flowing through your head. So, if you're ready, let's get started.

PART 1

Chapter 1: What is CBT?

None of us have control over every aspect of our lives. Therefore, we all must play the hands we are dealt as we navigate through life. Unfortunately, many of the circumstances we have to deal with are unpleasant and can leave us with lasting scars that can start a domino effect of negative thoughts that seem to never end.

It is important to understand that our ability to control many of the elements in our lives is out of our hands; we can't even control the thoughts that seem to flow endlessly through our minds. Still, this does not mean that we are completely helpless. While much of these things happen against our will, there are many things we can do to change the situations we face. At the very least, we can control how we respond to the external events and factors we deal with from day to day. We have the ability to control our reactions to events, thoughts, and internal beliefs about what is happening to us.

This adjusted kind of thinking is at the heart of CBT. It is a research-based approach to a treatment that will address a wide variety of mood and anxiety disorders. A fundamental belief behind CBT is that after experiencing certain situations, certain thoughts are developed in the mind. These thoughts lead to a variety of different feelings (many unpleasant), which can trigger negative reactions. Since feelings are not that easy to change, the concentration of the therapy is on challenging the thoughts that trigger those feelings and once you can control your feelings, you will be able to control your actions. As you develop these new skills, you will be able to manage your life better, which by extension, will allow you to build up a more positive view of the world, and thus have an improved pattern of behavior.

So, rather than pouring out your past to a therapist, opening up painful wounds of the past, and examining every minute detail of them, you and your therapist will work together to discuss the problems you are facing today, set a series of goals, and then apply different strategies and exercises you can apply to help you achieve them.

The Science Behind It

Unlike psychoanalysis, a common form of therapy developed by Sigmund Freud, where treatment is based on early life experiences and uncovering hidden memories that may have been buried, CBT takes an entirely different approach in healing the mind. In CBT, the problems are the foundation for the treatment.

In the early part of the 20th century when psychoanalysis was the accepted form of treatment, you would have spent an endless number of sessions talking about your past. The psychoanalyst would likely interrupt your dialogue with occasional questions about what you might think this image or that person meant, forcing you to think deeper about your reaction to certain events, people, and influences in your life. His goal would have been to get you to explore hidden messages stashed somewhere in your psyche.

At some point, the treatment would gradually take effect and your feelings would start to experience a change. This form of treatment, while effective, would often require a long-term commitment, which could take years to complete, if ever.

About the middle of the 20th century, we began to understand better how the brain worked. Initial research into the function of this organ started focusing on the more recent scientific discoveries relating to both animal behavior and then later in the field of metacognition.

As scientific research progressed over the last century, they have been able to develop a better understanding of learning, behavior, and cognition (our thoughts); it was a natural progression from Freud's form of therapy to what we use now. First, experimenters began to recognize that even animals could associate two separate events together. For example, an experimenter would ring a bell and then give food to a dog. After this action was repeated several times, the dog would automatically associate the bell with food. It was apparent that the dog understood that the sound of the bell meant food.

The next progression was the discovery that helped us understand how human behavior developed. Scientists wanted to understand what would make a person more likely to do one thing and less likely to do another. It was quickly understood that if you punish someone for an action (giving them an unpleasant result), then they would cease the action. The opposite was true with a reward system. If a reward was given, the individual would then be encouraged to continue the action.

This knowledge was quickly applied to mental health where they could now focus their attention on treating the behavior of a patient rather than rehashing all of its past experiences. The patients that are most benefited from this type of therapy are, in a sense, stuck in a certain pattern of behavior. The trigger for that behavior may have happened in his or her childhood or it may have happened in the past few months. The time of the trigger or the real nature of it is not as important as addressing the present state of affairs. The therapy, therefore, is focused on changing that behavior so that he can get out of his rut and start living his life again.

The next evolution of CBT was developed much later in the 60s and 70s. The focus then was on identifying what was really triggering that behavior. It was determined that it was the power of the thoughts and their effect on emotions, which in turn, was triggering the actions. Psychiatrists recognized that to understand how a person feels, they must first understand what a person was really thinking. If a person is struggling with anxiety, then chances are their thoughts were full of thoughts of danger or risk.

Up until that point, they had recognized the cause and effect scenario that was playing out in a person's mind. For example, someone who is afraid of elevators would see the elevator and automatically feel fear.

What they realized later was that there was one missing element that needed to be understood. It wasn't the elevator that was causing the fear in the person, but it was their mind's interpretation of what that elevator actually meant. This goes back to the original premise that what a person thinks is what dictates his emotions and his behavior.

However, we all know that what we think and believe is not always accurate. According to his book *Healing the Addicted Brain*, Dr. Harold Urschel explains that once you begin to analyze your thoughts, you will inevitably find inaccuracies in your reasoning. These inaccuracies tend to reappear over and over again in your mind until corrected. In other words, all of us have distorted thinking and need to be aware of them. In fact, just the simple act of recognizing and acknowledging them will automatically make it easier to change your thought process and help you to replace them with healthier thoughts.

So, if one can get inside the thought process and find the flaw, correcting certain behaviors could be addressed simply by correcting the wrong thoughts that were playing out in their mind at the time. This means that if a person was struggling with irrational thoughts, these could be replaced with more realistic beliefs through a process of challenging them and arming them with a new way of thinking. Analysts were no longer focused on finding the root cause of a problem and could now focus on changing the thoughts that trigger unwanted behaviors.

Through CBT, analysts could help their patients by giving them a set of skills that could be learned, practiced, and through a regular application on their own, could work independently to address their issues.

How it Works

As you read these words, the whole process sounds relatively simple. At first glance, you might have even wondered why a person would need a therapist to figure this out, and you'd be right. But, for those who are stuck in that endless loop of unwanted behavior, it still remains a real challenge.

It is a huge step from Freudian therapy to CBT. A person has to first come to the understanding that three core elements of our human makeup are completely interconnected: our thoughts, feelings, and behaviors. It is literally impossible to change one without changing either of the other two.

This concept can be illustrated simply. If you are struggling with feelings of anxiety, it is usually because your mind is overwhelmed with thoughts of dangers that you want to avoid. However, if your thoughts of danger are distorted, then your reaction to those thoughts will be strange, out of the ordinary, and even extreme. It is only a short step from a fear of danger to trigger our fight-or-flight response, which will either cause us to avoid the situation altogether or to take an aggressive stance.

The Basic Principles of CBT: In order for CBT to be effective, there has to be a good relationship between the therapist and the client. At the very least, there must be a mutual respect for each other and it is extremely important for the therapist to get a good understanding of how the client sees things.

From this area of mutual respect, the foundation for treatment can be laid. After that, there are several other basic principles that must be followed:

- CBT is not an open-ended therapy session – it has a time limit. In most cases, the time with a therapist runs only 10-15 sessions. This puts a slight pressure on the client to not procrastinate in getting started with the specific steps the therapist will give him as homework. He is less likely to put it off until a more convenient time.

- CBT is based on more than a century of research studies. From the beginning, a therapist can give a pretty reasonable estimate as to how long a program can run given the client's unique mental state of mind, and how much of a benefit he can expect to achieve in that time period. As they work together through the sessions, it will be easy to see which parts of the program are working well and which ones are not and make adjustments accordingly.

- CBT always has a very specific objective for the client. It is not about looking at the past but is more about achieving a specific goal. You can't make progress if you don't have a direction to go in, and it won't take long before you will see evidence that the program is helping you to get to the behavior you want to display.

- CBT is a team effort. The therapist doesn't "fix" you. You work together to address and discuss out a certain issue. Each party in the team has something to contribute; the specialist understands just how CBT works, and you (the client) understand how you work. Bringing the two together, as a team, allows both to tailor a treatment that will address the specific needs of the client.

- CBT is well-organized. CBT sessions are not hours of aimless ramblings with a therapist. Instead, they have structured in such a way that the patient knows exactly what his objectives are and the steps they are going to take to get there. The first sessions are dedicated to creating a roadmap outlining every step in the treatment plan. Throughout the program, both the therapist and patient can check to see if they are still on course or if adjustments are needed.

- CBT is about the present. While CBT looks at past events in a patient's life, its primary focus is on what can be done today to fix the problem. It stresses the importance of finding ways to change thoughts and behaviors today so that there can be immediate relief.

- CBT involves action. Both the therapist and the patient are fully engaged in the process. Without a commitment from both parties, it won't be successful. From the very beginning, it requires both to be fully involved with each stage of the treatment.

- CBT focuses on skills. Patients go through each therapy session learning different skills they can continue to use once the session is over. These skills teach you how to identify the mind games you play with yourself, the self-sabotaging strategies you use, and teach you how to stop those bad habits in their tracks and choose healthier and more productive options.

- CBT stresses practice. You have 168 hours in a week and can count on only one hour of time with your therapist. So, in order to get the most out of each session, it is important that you practice each of the strategies learned in your session on your own. The more work is done in between sessions the faster the progress will be.

Changes Thought Patterns and Belief Systems

Since Cognitive Behavioral Therapy is based on the evidence that our *perceptions* of the situations in our lives dictate how we feel emotionally, our perceptions of the world are, therefore, the direct result of what we think is really happening or will happen. These are the elements of our lives that need to be adjusted.

For example, an individual may find something in this book and think it is exactly what he is looking for. Because he *believes* it is beneficial for him, he then feels good and perhaps even relieved to have found the solution to his problem. However, another person may read the same information and feel entirely different. He may believe that it isn't the right solution for him or that following the exercises are beyond the scope of his ability. As a result, he will feel disappointed and maybe even discouraged at not finding the answer to his problem.

In theory, it is not the experience that causes people to feel the way they do, but it is their perception of the situation or their thoughts about the experience that needs to be modified. We all understand that when a person is under stress, their viewpoint of a situation is often extreme, inaccurate, and unrealistic, so if it is the perception of an experience that triggers negative habits, then this is the area that needs to be focused on.

With CBT, people are helped to identify those negative thoughts, give an honest and realistic appraisal of them, and then learn strategies to help them to readjust their distorted thinking. Once they are able to think more realistically, they will begin to see things differently and it will be easier for them to adjust their behaviors. The therapy works within a narrow focus of solving their immediate problems and making behavioral changes.

Its Purpose – Who it Can Help

Of course, CBT is not a catch-all for all psychological disorders. It has proven to be most effective for the treatment of anxiety, depression, and other forms of mood disorders including obsessive-compulsive disorder, bipolar, and PTSD. While the National Institute for Health and Care Excellence (NICE) specifically recommends CBT for the treatment of anxiety and depression, there are many more mental health issues that it can be very effective with.

Because of its unique flexibility, the therapy can easily be adapted to a wide range of needs. There is a whole list of psychological issues that can be benefited by it including:

- ADD/ADHD
- Alcohol abuse
- Anger management
- Anxiety
- Bipolar disorder
- Borderline personality disorder
- Chronic fatigue syndrome
- Chronic pain
- Depression
- Drug abuse
- Eating disorders
- Facial tics
- Insomnia
- Obsessive-compulsive disorder

- Phobias
- Post-traumatic stress disorder
- Psychosis
- Relationship problems (individual or couple's therapy)
- Schizophrenia

In most cases, it can be a very effective form of treatment without the use of medication, but in extreme cases, where the symptoms a patient is experiencing are severe, a doctor may suggest the use of medications to make it easier to get into a mental state where patients can effectively apply the strategies suggested as they go through the program.

How it Works on Anxiety and Depression

When dealing with anxiety disorders, one must first come to understand the true nature of anxiety. While unpleasant in many ways, a certain level of anxiety is necessary for life. Without it, we would find ourselves in an even worse predicament. Consider all the ways a healthy sense of anxiety can be of benefit.

Both anxiety and fear are natural forms of emotions. They serve as your body's personal alarm system and can occur as a warning sign that there may be some kind of risk or harm present. Fear, a heightened sense of anxiety, appears when a person is literally faced with a dangerous situation. Anxiety occurs when they "expect" that something unpleasant is about to happen.

A good example of this is taking a ride on a roller coaster. The feeling you get as the car is slowly pulled up that first big hill. You are in anticipation of what you know is about to happen as soon as you reach the top. Fear is the feeling you get when you are plummeting down the hill.

Both emotions are warnings and can trigger all sorts of bodily sensations. They tap into our flight-or-fight response so that we can quickly respond to those warnings, allowing us to flee, freeze, or fight. These instinctive emotions have been a part of us for as long as the human race has existed. It is a highly developed system that allows us to react quickly without having to put forth a lot of effort. They make it possible for us to have an automatic response.

The problem with these emotions is evident when they become overactive. It is natural for humans to have a very active imagination that allows our minds to create certain scenarios that we may encounter. For example, you may be anticipating how a certain job interview will play out. Your mind may conjure up something good or something very bad, both could result in anxiety even though there is no assurance that what you imagined will actually happen. This is evidence that our internal alarm system can be triggered even when there is no real threat to deal with.

The natural progression is that if our automatic thoughts are negative, we will respond accordingly. As a result, if we have conjured up images of a bad interview, our reactions could be so strong that we can take some very negative actions that could harm us in some way. For example, we might refuse to show up for the interview, we might choose to apply for a job that we find less challenging or risky, or we might become so flustered that we blow the interview altogether causing our worst fears to come true.

These choices could actually have a very negative effect on our lives causing us even more harm in the future.

Anxiety and depressive disorders can appear in a variety of forms. Episodes may be occasional or they may be constant. If these automatic thoughts are not addressed, they will only become more sensitive starting a spiral effect that will send a person deeper and deeper into a level of anxiety that will be even more difficult to get out of.

CBT is most effective because it works to correct those automatic thoughts. Those spontaneous thoughts we all have that spring to mind without prompting. Automatic thoughts start before we are born and will continue unabated until we take our last breath. It is our brain's way of processing the millions of bits of information it is constantly receiving from our five senses: sight, sound, smell, taste, and touch. If the brain receives any input from these sources under a stressful situation, it can create unfounded assumptions about a given situation or possible outcome. These negative assumptions create an unhealthy internal dialog that could build up to a point that it prevents an individual from progressing when faced with similar circumstances. Unless this internal dialog is corrected, the resulting negative emotions or behavior could take control causing the person to be paralyzed to the point of inaction or to exercise complete avoidance.

So, how can you tell if you have normal anxiety or have a disorder? Mental health professionals often refer to the *Diagnostic and Statistical Manual of Mental Disorders (DSM-5)* for the guidelines for making a diagnosis. An anxiety disorder is diagnosed when:

- The anxiety one experiences are excessive: being afraid of snakes may be normal but being petrified of worms would be considered extreme.

- The anxiety is constant and can last for weeks or months at a time: there are different time frames for each type of disorder. For example, when panic disorder symptoms last for a month or more, it can be diagnosed as an anxiety disorder. However, for generalized anxiety disorder, the symptoms must be present for a minimum of six months.

- The anxiety itself causes more anxiety: when a person becomes really upset by the mere fact that they are experiencing anxiety.

- The anxiety interferes with a person's normal activities: when you can't walk outside of your home, go into a public place, or can't work on a job because of the heightened level of anxiety.

There are several different types of anxieties that could be considered an anxiety disorder. They are very different from each other, so before treatment can be started, one must understand exactly what type of disorder they have.

Phobias: Phobias are powerful and irrational fears of a particular object or situation. Common phobias could be a fear of anything from snakes to something that is part of the natural environment. Most common forms of phobias might be an extreme fear of storms, flying, elevators, spiders, and even people. In many cases, these start with a traumatic experience, but not always. Sometimes, the actual cause of a phobia is never discovered.

Social Anxiety Disorder (SAD): These are fears of social situations. In these types of cases, the fear is often confined to the fear of embarrassment. SAD is different from other phobias because it often involves the mind "guessing" about what is going to happen or what someone else might be thinking. With other phobias, we may know that the thing we're imagining can actually happen. We know about dog and snake bites; there is plenty evidence of the consequences of such events, but social phobias are more likely related to what one might be thinking rather than as a result of what they believe.

Panic Disorders: Panic disorders usually come about without warning. Many often mistake a panic attack for a disorder, but panic attacks that might be experienced are usually the symptom of the fear or phobia of a certain situation.

The attacks come on suddenly like a loud alarm going off in the body. The sympathetic nervous system instantly triggers a "fight-or-flight" response that is very physical. It releases a huge dose of adrenaline into the system to help prepare the body for danger. As a result, the person will experience:

- Rapid heartbeat
- Faster and deeper breathing
- Dizziness
- Profuse sweating
- Digestive problems
- Tense muscles

A person may experience some of these symptoms or all, but the concept here is that the symptoms are not just a little change but are quite extreme.

Agoraphobia: Interestingly enough, agoraphobia is not a very specific phobia but is the fear of being in places where you feel it would be bad to have a panic attack. This causes the person to avoid public areas such as movie theaters, shopping centers, public transportation, etc. If they go anywhere at all, it is usually in the company of another person they feel is safe and would be able to help them if something were to happen. In extreme cases, the patient is reluctant to leave the house at all and may squirrel away inside for years at a time.

General Anxiety Disorder (GAD): This disorder presents itself as a persistent and pervasive sense of worry. It is not the normal concerns of everyday life that most people have but is so intrusive that it interrupts sleep, keeps the individual from concentrating which leaves them to feel exhausted all the time. Unlike other anxiety disorders, GAD is also evidenced by anxiety and worries that are spread out over a number of different areas. It is not specific to a single type of fear, thus the name of General Anxiety Disorder. The individual's life is plagued by a never-ending worry about "what ifs" to the point that they can no longer function normally.

People who suffer from panic attacks, phobias, obsessive thoughts, and a never-ending stream of worries about everything will gain the most from CBT. This type of therapy, unlike medications, treats much more than just the symptoms that you may have. It can help to reveal the underlying causes of those incessant worries and fears. As a result, patients learn to relax and see each situation in a completely fresh and less intimidating way. They become better at coping and learn how to think through all of their problems.

There are many different types of anxiety disorders, so each session will be tailored to the specific problem of each patient. For example, for those who may be struggling with obsessive-compulsive disorder (OCD), their treatment will differ greatly from someone who is struggling with an anxiety attack.

The therapist will work with them to first identify the negative thought and then teach them through guided sessions on how to challenge those thoughts. For example:

Negative thought: What if I have a panic attack on the subway? I'll pass out!

Distortion: Believing that only the worst can happen.

Challenge: Have you ever passed out on the subway before? It's unlikely that you will pass out on the subway.

Negative thought: If I pass out, terrible things will happen.

Distortion: Expectations clearly out of proportion with reality.

Challenge: If I do pass out, it'll only be for a few seconds. That's not so bad.

Negative thought: People will think I'm crazy.

Distortion: Drawing conclusions you know nothing about.

Challenge: People are more likely to show concern and help.

In the beginning, these thought challenges will be guided with the therapist, but in time, the patient will learn to do these themselves. For milder cases, an individual can work these out for themselves with the use of a CBT workbook. Negative thinking doesn't just happen all on its own but is an accumulation of a lifetime of negative influences. While it sounds simple here, it can be quite difficult to remold one's thinking into something that is more positive. Those suffering from anxiety and depression disorders will:

- Learn to identify anxiety and how it feels in the body
- Learn skills to help them to cope and relax to counteract the anxiety
- Learn to confront their fears (real or not)

Of all the psychological disorders you can find, anxiety is by far the most common.

- 18% of anxiety disorders are phobias
- 13% are social anxiety disorders
- 9% are general anxiety disorders
- 7% are panic disorders
- 4% are agoraphobia

Depressive Disorders

Depressive disorders come in a different form. Many may start with some type of injury that renders them unable to function for a time. This triggers a train of losses or disappointments of many of the things the person may use to enjoy. The loss of these once fulfilling activities starts him on a downward spiral where he begins to believe that he doesn't deserve them or that he's done something that has triggered the loss.

A depressive disorder can make it difficult to function in any capacity. They may feel like everything they do is ten times harder than it used to be. It could feel like trying to pull themselves out of a quicksand and they may feel that they are so pathetic or worthless that it is not worth putting up the effort.

Most people only know depression as a feeling of sadness, but its symptoms can vary widely depending on the type of depression they are experiencing. It can present itself in many forms. In many cases, the patient has no idea that he or she is dealing with depression, because often, the symptoms do not present themselves as they would expect.

Major Depressive Disorder: Another term for someone who is "clinically depressed" is a major depressive disorder. This can be described as a person who feels low for the better part of the day. However, while this is a very common symptom, it is not necessary for them to feel an overwhelming sense of sadness to be considered depressed. It could be evident in their lack of interest in any activities, many of which they may have used to enjoy.

At the same time, they may sleep excessively, eat more than usual (whether they are hungry or not), or stop eating altogether. They feel exhausted and have a hard time concentrating on what they need to do or even making decisions.

While we all have a tendency to feel bad about ourselves at one time or another, those who are clinically depressed feel so bad about themselves that they believe they are completely worthless and are at a very high risk of suicidal tendencies, so they need to be watched carefully.

According to DSM-5, there are nine different symptoms of depression; five are needed to be observed in order to be diagnosed as clinically depressed. Because of this, the condition of depression may look very different from one person to the next.

Persistent Depressive Disorder: There are ebbs and flows of a major depressive disorder, where the individual may feel up and good at times but at other times, they may struggle to even get out of bed. With a persistent depressive disorder, however, the symptoms may seem to be more chronic. To be diagnosed with a persistent depressive disorder, they have to experience symptoms consistently for a minimum of two years before they are diagnosed. In addition, they need to display at least two more symptoms listed in the DSM-5.

With a PDD, the symptoms can be milder, but its negative effects can be extreme. While the symptoms may not appear to be as severe as an MDD, one can never assume that its effects are not as bad. At times, the symptoms can be even more destructive because of their chronic nature, as the individual never gets any sort of relief from the pervasive symptoms.

Premenstrual Dysphoric Disorder: This type of depression usually appears just before and for several days into a woman's menstrual period. It is important to understand that this is not the same thing as PMS (premenstrual syndrome). The condition is much more extreme with women displaying volatile mood swings and irritability. In addition to feeling anxiety, they can often show physical symptoms as well including breast tenderness and bloating. The symptoms should appear before or during most of her menstrual cycles for as much as a year before being accurately diagnosed.

Getting diagnosed with an anxiety or a mood disorder is not always easy. In fact, the process can be quite complex. Each form of depression has to have a number of different labels that identify exactly which type of depression the person may be dealing with and its nature. In addition to the main symptoms listed above, there may be seasonal depression, postpartum, a form of melancholy, single, and recurrent episodes. Whatever the case, there is a good chance that applying CBT techniques can help to manage all of these symptoms. One of the first things an individual needs to do is to get an official diagnosis. Once the diagnosis is completed, the next step is to sit down and work out how it is affecting you in order to determine the goals you want to work towards and a time frame in which to complete them in.

How CBT Helps With Substance Abuse

While CBT is most often used for anxiety and depressive disorders, there are many more mental health issues that it can also help with. It has been highly effective in preventing the relapse of problem drinking and drug addictions. These are common behavioral patterns that can be readjusted by redirecting the thought processes that trigger the abuse of many substances.

A key element of CBT is teaching the patient to anticipate situations and give them practical strategies to help them to cope effectively when these conditions arise. These strategies could include things like analyzing consequences before a decision is made to use the substance, self-monitoring themselves so they can identify cravings at the onset, and avoid situations where they might be at a higher risk of using.

By learning to develop strategies that will help them to better deal with cravings and those situations where they will find themselves more vulnerable, they can reduce and eventually eliminate their use of these substances

With the treatment for substance abuse, the objective of CBT is to:

- Help the individual to identify those situations where they are more likely to use alcohol or drugs
- Help them to develop strategies to avoid such situations
- And to identify common problems and triggers, which could cause them to abuse substances

This is done through two major components: function analysis and skills training.

With functional analysis, the therapist and the individual works to identify certain thoughts, feelings, and environments that historically have led to substance abuse. From this, they are able to determine the level of risk the patient is likely to face, or under what situations could expose them to a potential relapse.

CBT also will give them special insight into what started them on the path to substance abuse in the first place. Then they will give them coping strategies, so they are equipped to handle similar situations in the future.

With skills training, the individual will learn better coping skills. They will work on unlearning those old habits that took them in the wrong direction and replacing them with healthier habits that will be of benefit to them. In other words, they will be educated about how to change their thinking so that they are better able to deal with the varying situations that will trigger more episodes in the future.

To date, more than 24 different trials have been conducted with substance abusers of all sorts including those who are addicted to alcohol, cocaine, marijuana, opiates, and tobacco. In each of these studies, CBT has proven to be one of the most effective forms of treatment in helping to restructure their recovery. However, it is not the catchall answer for everyone. Because each person is different in the nature of their problems, everyone will respond to this type of treatment in a different way. Still, the results speak for themselves and if you or someone you care about is struggling with some form of substance abuse, there is a good chance that they will learn something and gain some benefit from incorporating CBT into their lives.

Chapter 2: How CBT Can Help With…

We live in unpredictable times and every one of us may be expected to address a number of issues at any given time. It can be a pretty stressful life for most of us. However, if someone is dealing with intrusive thoughts on top of the problems they face, it can become a complex maze that they may struggle desperately to get out of.

Dealing with any problem is not really the issue for these people. When a situation arises, the difficulty occurs with the manner in which he or she may address it. If their solution is based on distorted thinking, it can create a myriad of new problems in addition to creating a snowball effect. People who are plagued with unrealistic and intrusive thoughts are usually more reactionary rather than taking a proactive approach.

CBT's goal-oriented approach encourages a hands-on and more practical direction which teaches the patient that there is a means to an end. They do this by making a comparison between where the individual is at the start of the problem and then working out the best path to take to achieve the set goal. The ultimate goal is to change the thought pattern, which in turn, will change how they feel, and by extension, their behavior. In this way, CBT has been used to treat a wide variety of problems that range from sleep disorders to relationship issues.

On the surface, the logic of CBT is very simple: change one's negative thinking to positive thinking. However, this is not always easy. It requires that the person first learn to identify the negative thought processes they have. This requires them to take a hard look at themselves and give an honest appraisal of how they really think.

This may involve applying a little mindfulness and seeing their thoughts as a neutral person. When they see that their views are overly negative or overly positive. They will be compelled to learn to adopt a more balanced view of themselves, those around them, and their environment; viewing everything as a neutral person. This kind of adjustment takes time, as they may not fully realize just how far their thinking has strayed from the normal way of things.

Patients need to learn how to identify when their thoughts are triggering negative emotions and when to stop them before this happens. Then, they need to examine those thoughts honestly to determine if they are truly giving a fair and honest appraisal of the situation; how realistic they are and how damaging those thoughts are to them.

This helps them to avoid falling into thinking traps (negative ways of viewing things) so they don't fall into the problem situation in the first place. These mental traps could become major landmines that could quickly throw them into a world of trouble. These thinking traps are often negative words they would never say to others, but they say it to themselves.

Consider if you don't like people talking down to you, then why would you do the same thing to yourself? It is possible to hurt your own feelings and damage your self-esteem. One of the first traps they are taught to avoid is those that you do without realizing it. Through CBT, you learn to identify these habits, stop them, and then change that point of view.

Problem-Solving With Depression or Anxiety

Problem-solving therapy empowers people, so they are better equipped to change the circumstances of their lives. They learn to work through challenges and take a proactive approach to their problems. Through this type of program, patients learn how to manage these issues through several core components:

Addressing the Problem: There are many approaches that people can take when they deal with a problem. Some will naturally be more submissive in the decision-making phase, others will apply avoidance techniques, and others will be more aggressive. Whatever strategy one takes, the treatment will focus on the developing thoughts and attitudes one applies to solve their particular problems. It will work to identify certain weaknesses and address those with cognitive steps and techniques.

Defining the Problem: The next step is to define exactly what the problem is so they have a clear understanding of what they are dealing with. For example, a client may see that they are constantly under stress while at work. The first conclusion might be that they are dealing with anxiety. However, other factors may be at the root of the anxiety. They may not be assertive enough on the job and as a result, are allowing people to overburden them with additional work. They may not be setting boundaries, so people do not know their limits, and this is adding to their stress level.

Developing Strategies to Help Them Cope: Together, they work with the therapist to outline very specific steps that will help them to change their behavior and provide a number of possible solutions to the problem. By taking into consideration a wide variety of possibilities that will work, they may feel more empowered and actually begin to believe that they can work it out.

Execution: Finally, when they have broken down their goals into smaller easy-to-achieve steps, they are better capable of taking the necessary action to solve the problem. As they progress through the different steps, they will gain confidence and are more likely to follow the plan and successfully solve their problems.

There have been many studies that have definitively shown that this type of therapy can be a very useful form of therapy on its own. However, it can be even more effective when it is included in a full CBT treatment plan, yielding even better results.

CBT with Sleep Disorders

Often, people who struggle with sleep disorders have an extremely difficult time dealing with life in general. With a lack of sleep, they will naturally suffer from low energy, and when their energy is low, they are likely to also have poor nutrition. Poor nutrition contributes to brain fogginess, and in the end, they are caught up in a vicious cycle that is difficult for them to break free from. CBT can help them to get to the root cause of their sleep disorder and correct the wayward thoughts that are interrupting their ability to sleep.

Insomnia, a common sleep disorder that affects many people, has been treated very effectively through CBT. The condition makes it hard to fall asleep and even more difficult to stay asleep once you do.

CBT-1, the form of therapy used for sleep disorders, is a structured program that helps you to replace those intrusive thoughts with the kind of habits that will be more conducive to getting good sleep. The right treatment used for your sleep disorder will depend on a variety of factors.

Stimulus Control Therapy: This form of therapy focuses on removing the factors that may be contributing to your inability to sleep. These could include setting up routines that will support a better bedtime routine, one that is more conducive to sleep. It may involve setting up a consistent time to go to bed and wake up, avoiding naps, and restricting the use of the bed for anything other than sleeping and sex. No eating, watching TV or talking on the phone while in bed.

Sleep Restriction: This treatment would prevent you from being in bed for too long a time without sleeping. If you're in bed for more than 20 minutes without sleeping, it can add to your sleep deprivation. With sleep restriction, the amount of time you spend in the bed is shortened to only when you can sleep. Once you have developed a healthy sleep period, this time can be extended until you reach a point where you can sleep through the night.

Sleep Hygiene: This will get rid of those negative habits that interrupt sleep. Smoking, too much caffeine late in the day, consuming too much alcohol, or lack of exercise can all be part of the problem. CBT therapy can help you to reduce or eliminate those habits so that you can more easily relax before bedtime.

Sleep Environment Improvement: This treatment focuses on habits that create a positive sleep environment. Strategies to keep the room dark, quiet, and at the right temperature can be very effective in teaching a patient to relax.

Passive Awakeness: This focuses on avoiding any attempt to fall asleep. Interestingly enough, worrying about not being able to sleep can contribute to one's inability to sleep. Learning how to not worry about it allows the body to relax so you can drift off to sleep more easily.

Biofeedback: Once a person is aware of how his biological functions are operating, they can learn how to manipulate them. Many of these factors can interrupt one's sleep. A person's heart rate, tension in their muscles, or body temperature can all interfere with their regular sleep patterns. Learning how to mentally adjust these and other biological elements can put your body into a more comfortable state where sleep will come easily.

Whether your sleep disorder is mild or extreme, using CBT-1 can help all sorts of problems. Even if the cause of the problem is physical, as in cases of chronic pain, or mental, as in cases of anxiety and depression, the problems tend to go away once the right kind of therapy is undertaken, all without negative side effects. It will require consistent practice and applications, but the longer you stick with it, the better the end results will be.

Chapter 3: Common Causes of Mood Disorders

When you first start Cognitive Behavioral Therapy, one of the first things you will have to do is to identify the root cause of your dysfunction. In nearly every case, the problem lies in your automatic thoughts. These are at the heart of the theory behind this type of treatment.

Automatic thoughts are known to pop into your mind without any effort. They are the brain's means of processing all of the information it is receiving from the environment. These thoughts do not represent any type of fact about the situation that it is processing, which is why they can cause so much damage. The type of automatic thoughts that appear immediately after dealing with a specific situation can be referred to as your "instinctive" response. They appear so quickly that there is literally no time to reason on the situation or apply logic, so relying on them to make decisions is often the reason why a person would display dysfunctional behavior.

In your first session with the therapist, you will learn to recognize your dysfunctional thoughts. Likely, you will backtrack from a specific experience in order to identify the actual thought that triggered your negative emotion. You should pay close attention to any thoughts that cause you to have a reflex change in mood in response to it. These are the dangerous thoughts that are connected to what you truly believe about something.

A good example of this may be how you react when you watch someone speaking in public. Your knee-jerk reaction to what is said is a reflection of what you truly believe about that situation. If there is a great deal of approval from the audience but you don't like it, your mind won't take the time to contemplate his words. If you are applying your feelings to yourself, your thoughts might reflect a general negativity about yourself. "People will never respond to me like that," or "I wish I was as happy as he is." These personal negative reflections are labeled as "dysfunctional automatic thoughts." These beliefs will trigger equally negative feelings that are, in turn, triggering certain behaviors you want to change.

To address these thoughts through CBT, you will first have to figure out if your negative thought patterns are caused by a physical imbalance in your system. There are several things that could tell if you are physically out of balance.

Chemical Imbalances

There is quite a bit of dispute about whether chemical imbalances are a real cause of mental disorders or not. It was once believed that chemical imbalances are the result of too much of some chemicals or too little being produced in the brain. These chemicals, called neurotransmitters, are used to aid in the communication between the neurons of the brain. You may have heard of many of them: dopamine, serotonin, and norepinephrine are just a few. It was generally understood in the world of psychiatry that mental disorders like depression and anxiety are usually the result of these chemicals being out of balance in the brain.

These types of disorders can be extremely complex, and finding a definitive answer to your problem is not always easy. However, a good way to assess whether or not you're dealing with these types of imbalances is by looking at the symptoms.

- Feeling sad, worthless, or empty
- Tendency to overeat
- Loss of appetite
- Inability to sleep
- Sleeping too much
- Feeling restless
- Irritable
- A constant feeling of dread
- Listless
- Not wanting to associate with others
- Lacking empathy
- Feeling numb all over
- Extreme mood swings
- Inability to concentrate
- Suicidal thoughts
- Thoughts of hurting others
- Inability to perform normal everyday activities
- Hearing voices that are not there
- Substance abuse

While we don't know the exact cause of these imbalances, many researchers believe that it could be a combination of genetics, our environment, and social influences on our lives. It is even unclear how these are able to affect us and create mental disorders. What we do know is that even if you do have a chemical imbalance, it is not the end of the situation. By applying the techniques of CBT, it is possible to change to more positive behaviors, which can, in time, correct them.

Dopamine: The brain uses dopamine to control both your movements and your emotional responses. When it is balanced properly, it is an essential component for physical and mental wellbeing. It also helps facilitate important brain functions like your mood, your ability to sleep, how you learn, concentration, motor control, and even the ability to remember things. If you are low in dopamine, you are likely to feel:

- Muscle cramps
- Tremors
- Aches and pains
- Stiffness in the muscles
- Have difficulty keep your balance
- Constipation
- Trouble eating or swallowing
- An unexplainable weight loss or weight gain
- Gastroesophageal reflux disease
- Trouble sleeping
- Low energy
- Poor concentration
- Excessive fatigue
- Unmotivated
- Extreme sadness or hopelessness
- Poor self-esteem

- Anxiety
- Suicidal thoughts
- Low libido
- A lack of self-awareness

A dopamine deficiency may be linked to a number of mental health disorders. It may not be the direct cause, but it can certainly contribute to the severity of the problem.

Serotonin: While your body mostly uses serotonin in the digestive system, it can have an effect on every part of your body including your emotions. It can also be found in the blood platelets and in various parts of the central nervous system. It is made from tryptophan, an amino acid found in common foods like nuts, cheese, and red meat. When you have a deficiency of serotonin, it can trigger anxiety or depressive mood disorders.

Because it can be found throughout the body, it can have a major impact on everything from your emotions to your motor skills. It is the body's natural way of keeping the mood stable. It also helps with eating, digesting foods, and allowing us to relax enough so we can sleep. When you have a good balance of serotonin, your body can:

- Fight off depression
- Control levels of anxiety
- Heal wounds better
- Maintain good bone health

You will also find that you are:

- Happier

- Emotionally stable
- Calmer
- Less anxious
- Able to concentrate better

If your serotonin levels are too high, it could be a sign of carcinoid syndrome, a condition that causes tumors in the small intestine, bronchial tubes, appendix, or the colon. Your doctor could determine your serotonin levels through a simple blood test. Symptoms of a serotonin deficiency are:

- Poor memory
- Signs of aggression
- Low or depressed mood
- Anxiety
- Cravings for sweet or starchy foods
- Insomnia
- Low self-esteem

You are equally at risk of major health problems if your serotonin levels are too high or too low, so it is very important to keep a good balance to protect not just your physical health but your mental health as well.

Other Imbalances: There are other chemicals in the brain that can have an effect on your moods and emotions. However, as we now understand how the brain works, it is important to realize that there is still an ongoing debate as to what extent these imbalances actually cause or affect the moods and emotions.

It should be a foregone conclusion that no person is born with these anxiety disorders. Whether the imbalances are a result of one's genetic make-up or are from the environment, it is clear that there is only one way that we know of to fix it. We have already learned that habits and behaviors are developed by using the neural pathways in the brain to send signals back and forth. Since these chemicals are responsible for facilitating that communication, it only stands to reason that the best way to treat imbalances is to strengthen those connections.

Because it focuses on those automatic thoughts, even if there is a chemical imbalance, it can be reversed by regular practice of the routine steps and strategies a patient will learn while working through the steps of CBT.

Chapter 4: Understanding Your Moods and the Way You Think

Understanding your mood and the way you think is not easy. As you analyze a particular situation where you have been feeling those negative emotions, you have to start with a serious self-analysis. Your goal is to figure out exactly what kind of thoughts and feelings you have lying just underneath the surface. You can do this by asking yourself some very pointed questions:

- What was I doing when I began to feel that way?
- Where was I?
- Is this the only place where I feel like this?
- What was my behavior like before and after this episode?
- What hidden beliefs do I have that are showing right now?
- What causes me to intensify those feelings?
- Who was I with?
- Am I the same with everyone or just certain people?

An honest answer to these types of questions can be very revealing. For example, a person may start to feel depressed when he is alone and away from other people. These types of questions help you to get to exactly why you have experienced those sudden mood changes. If you're a person that suddenly feels awkward and ashamed when you're in the gym, you may be harboring hidden beliefs in relation to your body image. If you start to feel negative emotions while out in public, it could mean that you have serious concerns about your abilities to perform in front of strangers.

After doing this type of exercise, most people are surprised when they learn so much about themselves this way. It can be alarming to discover that their automatic thoughts, which are often just fleeting in the mind, can have such power, triggering emotions unexpectedly as if they just came on without warning. Usually, these thoughts reveal a very specific issue and once you know what it is, you can now direct your energy to that particular thought process.

How to Diagnose Your Negative Thought Patterns

As you get started working your way through this stage of therapy, you must realize that no two people will have the exact same disorder. While they may have been diagnosed with the same label, your personal experience will be very different from someone else, because you bring to it your own personal body of experiences.

Even finding the answers to questions like the ones listed above are not easy to do. You may have to get a frank viewpoint from someone close to you and whom you can trust to get the ball rolling. Some complete the entire first session without learning what is at the heart of their problem. However, they will have learned how to start trying to think differently about their experiences.

This is not to be surprising. Most of us have gotten pretty good at hiding some of those negatives that permeate our lives every day and have done so for years. It's not likely that everything is going to come out in the first wash. Still, if you persist in this type of self-analysis, there is a very good chance that you will be able to identify many of the tricks and tactics you have created to cover up your true feelings about certain things.

As you go through this process, don't be afraid to look at every aspect of your life and not just in the obvious areas. It is quite often that the real trigger has little or no direct connection to where your thinking went awry. For example, you may be having a communication problem with your spouse. He or she may think you are distant, uncommunicative, or just has lost interest in the relationship. However, if you're dealing with a hidden anxiety or depression that started many years prior, it may still be affecting you and causing you to separate from your partner.

Imagine it this way. You are a child and you have a close and dear friend who died of a sudden illness. Your relationship was strong, and you were always very happy when with that person. However, their death suddenly left a huge void in your life. On the surface, you were able to eventually get over it and move on, but that pain was never clearly addressed. Now, in a relationship, you are afraid to commit that way to another person.

This kind of hidden pain could be what is causing those automatic thoughts that are firing in your brain.

- I can't commit to another person, they will leave me.
- I can't go through losing someone again in my life.
- It's not worth it. Everybody is going to die anyway.

A past loss may have been the real reason you don't trust or are afraid to communicate with your spouse. People die, friends move away, jobs are lost, and a myriad of other conditions could be behind the reason for your automatic thoughts. If you were a child when it happened, it is quite possible your grief was either not recognized or considered unimportant in a world where everyone else is in grief too. However, being able to learn the hidden secrets that are triggering those automatic thoughts can give you great insight on where they come from and why they are resurfacing. As a result, you have a basis for addressing the issue now and correcting those thoughts from a more realistic position.

Now, this may seem strange in light of what we said earlier; that a CBT therapist does not reach back into our past to find the root cause of a problem. This is true, however, as you begin to recognize these automatic thoughts, it is only a small step to making the connection between your past and the present.

Once you are able to identify those thoughts, there are several ways you can deal with them.

- If you feel the thought is indicative of a larger issue, you can choose to put all your energies into correcting that thought.
- If you feel that there is another problem that better identifies the issue, you can choose to focus on the two thoughts together.
- If you discover that there are other issues that are beginning to surface, you can put the initial thought aside and focus on the bigger and most important issues first.

After you have uncovered these hidden thoughts, the next step is to analyze each thought and rate them as to how important they are. If one thought seems to produce intense feelings in one way or another, it is likely the one that has the biggest impact.

Sometimes, it can be difficult to separate the different thoughts in your mind. In that case, it is helpful to keep a notepad nearby and write down the thoughts as they appear. As you write down each thought, make sure to also take note of what was happening when it occurred and the feelings it evoked. If you do this over a period of time (perhaps a week), you'll begin to notice a definite pattern emerge.

There are several different categories of automatic thoughts, and by journaling them as they happen, the patterns that emerge will help you to classify the type of thoughts you're having. Below is a list of some of the most common beliefs that many people develop that can trigger negative emotions.

- All or nothing: A person who believes everything is either black or white with no middle ground or gray area.
- Catastrophizing: A belief that every situation will produce the worst results.
- Discounting Positives: The belief that all positive experiences are false.
- Emotional Reasoning: Allowing one's negative feelings about a situation to be the deciding voice in their head.
- Jumping to Conclusions: Automatically concluding negative results for a situation with any evidence to back up assumptions.

- Labeling: The act of giving negative labels to ourselves or to other people. Calling yourself a loser or a failure rather than making the effort to change the quality you don't like.

- Magnification/Minimization: Putting a lot of emphasis on anything bad and downgrading anything good. Really stressing out about a mistake but not willing to accept compliments or acknowledge achievements.

- Negative Bias: Seeing only the bad in any situation and dwelling on the negative, despite the fact that there are many positives.

- Overgeneralizations: Using a single negative experience to represent every other similar event.

- Personalization: Believing that negative comments or actions from others are about you or believing that you are the cause of a bad event even when you had no connection with it.

- Should/Must statements: Having expectations based on what you believe should be done. These often stem from distorted perceptions of what others may believe about you and are likely not in the realm of reality. This could cause you to feel guilty for not meeting these abstract standards or often excessively high expectations.

Likely, as you read through this list, some of these dysfunctional automatic thoughts popped out at you. All of us have them from time to time, but if you think they are interfering with the progression in your life, then these should be the ones that you should focus your CBT exercises on.

Intrusive Thoughts

In our minds, we have many thoughts throughout the course of a single day. It is estimated that each of us has between 50-70,000 thoughts in a single 24-hour period. Most of them, we are able to quickly dismiss as insignificant and unimportant. But there are those disturbing thoughts that seem to get stuck in our brain and no matter what we do; we can't seem to get rid of them. They make us feel sad, frightened, and sometimes, even sick and can create a great deal of turmoil in our emotions.

The fact that they are so intrusive and they seem to camp out in our head can cause us even more distress; some to the point where they interfere with our regular routines and activities causing us to feel ashamed, guilty, or afraid. Anyone suffering from an anxiety disorder like OCD or PTSD can quickly relate to the kind of damage an intrusive thought like these can do.

These thoughts are always unpleasant and can even make you feel repulsed. They can include acts of violence, inappropriate sexual acts, or extreme criminal behavior. Those in relation to anxiety could be an excessive worry about future events or threats. Strategies learned through CBT can help to reduce their frequency and even help to lessen the extremes that they may take. Addressing these feelings would be a good way to start getting rid of them.

Examples of Some Intrusive Thoughts

- Unwanted sexual fantasies involving a child, animal, or another person close to you

- Unwanted sexual thoughts involving someone you work with but are not really attracted to.
- Imagining yourself committing a violent criminal act
- Fear that you will say the wrong thing in public
- Doubts about your religion or thoughts of doing something forbidden
- Doubts about your inability to do well on an exam you know you have prepared for
- Recurring thoughts about getting a rare disease
- Fear of death
- Repeated memories about something humiliating that happened in your childhood
- Repeated memories about a violent experience you had in the past

It is important to know at this point, that having intrusive thoughts is not indicative of a disorder. Everyone has them at one point or another. According to the *Journal of Obsessive-Compulsive Disorders,* 94% of people in the world have intrusive thoughts. It is one of the most common mental activities that we all have. What is different in those who have anxiety or depressive disorders is our reaction to those thoughts.

When you are healthy and your mental state is balanced, you are able to dismiss those thoughts and they won't upset you. When you struggle with those thoughts, it is because you have already associated a great deal of importance to them, which is an indication that you have an internal belief that they are true and accurate. In such cases, your mind starts to create a full narrative about those thoughts and conjure up its own implications about what kind of behavior you should display or what your future actions should be.

To counteract these thoughts, one of the first things you need to do is analyze them, so you can convince yourself that they are NOT true. This is especially important when you are dealing with some of the above thoughts that may be violent or otherwise inappropriate. It does not mean that you really want to do those things. If you continue to accept these things as true, then it can saddle you with an immeasurable sense of guilt and shame, which could cause you even more problems.

If you're struggling with these types of intrusive thoughts, then take some of the following steps to apply some CBT-based strategies to combat them.

Identifying Your Triggers for Anxiety and Stress

It is one thing to know what your problem is and another thing entirely to identify what is triggering the symptoms you're experiencing. Finding these triggers can be quite complicated as it is most likely due to a combination of factors including environmental, physical, and genetics. Even so, there are some events or personal experiences that may trigger certain forms of anxiety or, at the very least, make them much worse.

So, whether you're dealing with symptoms of anxiety, depression, or stress as a result of a genetic factor that you have no control over, the condition can actually deepen as a result of certain events or experiences you are having in your day-to-day life. These events, emotions, and experiences are called triggers.

These triggers can vary from one person to the next. However, in most cases, you're likely responding to several triggers at one time. It may seem, on the surface, that your reaction to a given situation comes completely out of nowhere, but in most cases, there is always something there, just underneath the surface. So, while it may be uncomfortable to do, it is important to root out your personal triggers, so you can take proactive steps in managing them. Below are some of the most common triggers that many people have.

Health Issues: Often, your own health may be playing a part in increasing your anxiety or depression. Diseases such as cancer or another type of chronic illness can trigger anxiety or even make it worse. Just getting a diagnosis of a disease like cancer can immediately evoke powerfully negative emotions that can be very difficult to combat.

To counteract this kind of trigger, you need to take a proactive approach by working together with your doctor and even discussing it with a therapist as they can give you strategies to help you manage your emotions as you go through your treatment.

Medications: The intrusive thoughts and negative emotions may be the result of the medication you're taking. While there is always a risk with certain prescription drugs of generating negative thoughts, even over-the-counter medications have been known to trigger anxiety and/or depression. This is due to the active ingredients that are working on your system. If you are able to dismiss these thoughts as a result of the medication, then great, but many people do not realize that the feelings they generate can trigger a number of side effects that could increase your level of anxiety.

Common medications that have been known to trigger anxiety /depression:

- birth control pills
- cold/flu medications to treat a cough and congestion
- weight loss medications

While for most people the symptoms will be non-existent or minimal, for those who have anxiety and depression disorders, these could pose a real problem. If after taking any medication, you find the intrusive thoughts are increasing or are causing negative emotions, speak with your doctor and get him/her to help you find an acceptable alternative.

Caffeine: Believe it or not, many people are addicted to caffeine, not realizing that it can be a very strong trigger for anxiety. Research has shown that people who are already susceptible to panic disorders or social anxiety disorders will be even more sensitive to the negative effects of caffeine.

To combat this, one of the best things you can do is to reduce or eliminate the amount of caffeine you consume and substitute it with other options for the stimulation you need.

Your Diet: Believe it or not, your diet can be a major contributing factor to your mental stability. Skipping meals, especially, can have a negative impact on your mental state. When you miss a meal, it causes your blood sugar levels to drop, which can trigger anxiety along with a host of other symptoms including that jittery feeling, nervousness, or agitation.

The best way to counteract this problem is to eat regularly and make sure that you have a diet that is balanced so that you get an adequate amount of nutrients in your system. The old rule that you must eat 3-5 meals a day is no longer encouraged. However, you need to make sure that your diet consists of enough nutrients to sustain you on a daily basis to prevent a drop in blood sugar.

Negativity: Nothing happens in our body unless it first happens in our mind, including anxiety. Your brain is the control center for your entire body so when you have anxiety, your self-talk can have a major impact on how you function. If you tend to be negative in your conversation or you are known for using a lot of negativity in reference to yourself, it can create a perpetual negative outlook on everything you do. Learn to use more positive expressions in your conversation and your feelings will soon follow. If you find this difficult, consider working with a therapist for a while until you learn to change your automatic conversation to a more positive and up-building form of self-talk.

Financial Problems: Debt can be a painful burden to undergo. Constant worry about not having enough or finding ways to save money can trigger a great deal of anxiety. While you may not be able to change your circumstances immediately, learning to manage these types of triggers can help to ease a lot of your intrusive thoughts. Working with a therapist can help guide you to a process that could help to get your mind to relax more.

Public Situations: Often, public events that include a lot of strangers can make one feel uncomfortable. The pressure of having to interact feels like walking through unfamiliar territory and can trigger many anxious feelings.

Worries about such occasions can be reduced by asking a trusted friend to accompany you so you won't feel like you're on your own. If your anxiety about these things is particularly intense, consider learning some coping mechanisms from a professional therapist so you are better equipped to managing them.

Conflict: Conflict in any form can be very stressful. Whether it is a relationship problem or a disagreement at work, these can trigger some of the worse anxiety you can deal with. If this is becoming a particularly stressful area for you, consider learning some conflict resolution strategies to learn better ways to control your feelings when conflicts arise.

Stress: We all face stress on a daily basis, but some of us get a bigger dose of it than others. Daily stress, when it is intense and constant, can trigger extreme anxiety symptoms along with other health problems. When under stress, you are likely picking up other habits that will contribute to your anxiety levels. Habits like over drinking, skipping meals, or insomnia will only intensify the anxiety.

Learning to first identify these habits and a few coping mechanisms will help you to handle them better so that they aren't allowed to overwhelm you.

Public Events/Performances: Situations that require you to speak or perform in public can be extremely stressful. Whether it is speaking up at a business meeting or competing at a special event, the stress can trigger all sorts of negative emotions and behaviors.

Try working along with a trusted friend or relative with some positive reinforcement to help you prepare and feel more comfortable beforehand. Regular practice with people you trust and even having them along with you for support can help you to feel more confident and relaxed.

Your Own Personal Triggers: It may be that you have triggers that are unique to you. These may not be so easy to identify. They could be something as simple as a smell that triggers something in your mind, the sounds of a song, the resemblance of a place or any number of things that can bring back a memory of a past traumatic event. This is common with people who suffer from PTSD. The triggers could be any number of things in their environment.

If you can identify these triggers, you can then take the next step to address them. Since these personal triggers are not likely to be those experienced by others, you will have to do a little bit of soul searching to uncover them. To help you find them, try these tips:

- Start keeping a journal where you take note of when your anxiety or depression is heightened. Record your thoughts and your feelings that occur at the same time. For example, answer the 5 W's (who, what, when, where, and why). After a while, you'll be able to see a certain pattern begin to emerge that will give you clues to your triggers.

- Be honest with yourself. Negative self-talk stems from a poor assessment of your true values and qualities. As long as you stay within that negative mode, it will be difficult to uncover the triggers to your anxiety and depression. Don't just accept the first thing that comes in your head. Be patient and be willing to look a little deeper into your life (past and present) to identify how they are making you feel.

- If the above two tips don't work, consider spending some time with a therapist. Even with all the effort you put out, some triggers have been so well-hidden that no matter what you do, it will be hard to pull out. Working with a professional can make it easier and take a lot of the pressure off of you.

Once you know and have identified your triggers, you have accomplished the first step in Cognitive Behavioral Therapy. Now, you can make a list of the problems you're facing and choose the ones you'd like to address and you're ready to move on to the next step in CBT.

Chapter 5: How it Treats Mood Disorders

During your first week of CBT, after you've identified the major problems you want to address, you'll have to go into setting goals for yourself. You know the type of anxiety and depression you're experiencing. Now, you need to work on a system for addressing these issues.

Practical Applications of CBT

Since no two forms of anxiety or depressive disorders are the same, there is no single way to deal with the challenges that each individual may face. Even those with the same diagnosis will have a different approach to their treatment. This means that before setting your goals, you need to have a pretty clear idea of what your specific problem is and what challenges you are facing. Once your unique situation is understood, then you can determine what kinds of changes are needed.

Generally, this is done at the first meeting with the therapist. You will be asked a number of questions that, at first, may not seem too important. For example, the therapist is likely to ask, "*Why are you here?*" Or in the case of trying CBT through the pages of this book, ask yourself, "*Why are you reading this book?*" This will prompt you to do an internal examination of your motivation for wanting to try CBT.

You might also spend some time determining what your strengths are. While you are looking at CBT to solve many of your problems, you will use your strengths as a tool to combat your struggles. There are several ways you can uncover your strengths. These are usually the things that people tend to recognize and admire about you, those qualities that others appreciate about you. Maybe, at this point, you don't see these qualities in yourself, but others do, so consider talking to someone close to you to find out just what they see in you that can be useful.

Next, you will need to take a step back and examine how your life is going at the moment. But don't just do a surface look at the conditions. Take extra notice of how the anxiety/depression is affecting you. There are several different aspects that you can examine closely.

- **Relationships:** Relationships often affect our overall health probably more than anything else. When your marriage is on the skids, it usually means that you're not satisfied with your life in general and that attitude can lead to some very negative thoughts. However, if you have a life full of positive relationships and full of supportive people, your outlook on life will have a more positive spin.

  As you examine the relationships, look at how your anxiety/depression is affecting them. For example, you may have a supportive relationship, but is your depression causing you to pull away and not spend as much time with them. Your anxiety may be causing you to live your life on a short fuse, constantly snapping at them or keeping you angry and on edge all the time.

  Don't just look at a relationship with your spouse or significant other but consider how it is playing out with your parents, siblings, friends, children, and even coworkers. Determine what is good in your relationships and what is not. How do you feel about the people who have moved out of your life (either by moving away, a death, or an unresolved disagreement)?

Finally, think about your relationships that are affecting your anxiety/depression. Are they making you more anxious or more depressed? Take note of the answer to these questions because you will have to face them later.

- **Physical Health:** Next, take some time to analyze your overall health. Here, you want to consider your eating habits, the amount of physical activity you're involved in, and what substances you regularly use (alcohol, drugs, etc.) The goal here is to draw a line between your regular habits and how they are affecting you physically. These can easily affect your moods and cause anxiety or depression.

As you consider health, think about any chronic health issues you're dealing with. Conditions like high blood pressure, diabetes, asthma, or other ongoing physical challenges can create huge inroads on your mood.

If you're getting more physical exercise, how are you feeling about it? Do you enjoy your regular routine or do you find it a chore; something that you just have to get done? Is there a kind of activity that you enjoy more than others or do you just not like moving at all?

Look at how much alcohol or drugs you use each day. These include prescription drugs as well as less favorable ones. If you've had a problem with substance abuse in the past, this is something that you need to give some serious thought to.

The same weight should be given to your diet. If you are a stress eater, you are likely eating foods that are hampering your ability to function. You may not even realize that you're eating these foods. Another concern should be if you're eating enough or overeating. These habits will affect your weight, which could have a direct impact on your own self-image, self-esteem, and self-respect; all of which will affect your moods.

Sleep also can affect your physical health. If you're not getting enough or you're getting too much, it can have an impact on your daily routine. It can keep you from getting the kind of movement you need and even the kind of foods you eat. Look at any external factors that could be affecting your normal sleep routine: pets, partners, work, kids, etc.

All of us need to have some downtime. If we allow our lives to encroach on that time, our activity level will suffer. There are probably many things you'd like to be doing and many things you'd rather not be doing. If your activity level is too high, chances are that you haven't had much time to relax and destress. You may also not have had any time to do the kind of things you enjoy doing like sports, hobbies, travels, etc. You can be highly active and still not have time to do the things you enjoy.

You will take this week to examine your entire life to see how these factors and others are affecting your mood. Afterward, go back over everything you've noted and see how it affects you. First, look at how it has affected your life and your mood in the past, but then take some time to meditate on them. Note what negative emotions come out as you're doing this review. Common feelings that emerge just talking about these things include anxiety, joy, sorry, overwhelm, sad, melancholy, etc.

Setting Goals

With a better understanding of the problems you're coping with, it'll be much easier to set specific goals for you to start working towards. In CBT, goals need to be very specific, something that is measurable and within reach. You need to concentrate on the type of changes that will actually be meaningful to your life and that you feel you can reasonably achieve within a short amount of time.

It is important to not be vague at this point. Remember that CBT has a time limit. Whether the program you are on is six weeks or ten, your goals should reflect where you reasonably expect to be at that point. So, goals like "I want to be happier" or "I want to stop worrying so much," are not specific enough to maintain your motivation for the duration of the program.

Instead, your goals should have very defined parameters. "I want to be able to get up on time every day," or "I want to find time to do something I enjoy." Your goals need to relate directly to the specific problems you are facing. "I want to voice my opinion more in employee meetings."

You want to avoid making grand goals that will require a major commitment; instead, focus on something that is well within your ability to achieve in a one, two, or three-week period. When you make large goals that are far out of your scope of ability, try to set a series of smaller goals that will help you to reach the larger ones at a later date. You can have these larger expectations set for years later, but it's best to break them up into smaller steps with each one taking you closer and closer to what you ultimately want to achieve. As you progress through the program, you will develop a more balanced view of yourself, which will help you to improve in your ability to accomplish more.

Building Self-Esteem

When you are suffering from low self-esteem, CBT can be very practical in helping you to identify the source of your negative thoughts, and how it influences your behavior. By incorporating these strategies, those dealing with self-esteem issues can work their way towards a healthier lifestyle.

Because CBT works on your personal perception of yourself, the techniques learned will help you understand how your mind works to create its own meanings in your experiences, and then teach you how to reframe your negative views and alter them by focusing on building a more positive structure of thoughts to rely on.

Throughout your sessions, you may learn techniques like:

- **Cognitive Restructuring:** where you focus on thought patterns and their sources. By regular reflection on these patterns, you will learn how to analyze your method of judgment and start reshaping your perception into more realistic possibilities.

- **Behavioral Activation:** where you identify those situations that cause you discomfort or make you afraid. By engaging in this technique, you are encouraged to step out of your comfort zone and gradually learn how to insert yourself into new situations.

- **Assertiveness Training:** through assertiveness training, you learn to reclaim your self-confidence. Through this, you are motivated to step out of your area of comfort and assert yourself in various areas where you can build up your self-confidence.

- **Social Skills Training:** through this form of therapy, you learn to work on improving social skills. It may include learning how you interpret and analyze many of the social situations you may have to interact with. Since low self-esteem comes from a negative perception of interactions with others, learning how to interact in a variety of social environments will make you aware of how your mind processes these situations and redirects them into a more positive direction.

Anyone suffering from a low self-esteem can learn how their thoughts relate to their actions and how to reflect on their impact and work to redirect them in a healthier and more positive lifestyle.

Verbal Self-Talk

We talk to ourselves all the time, but when you have negative self-talk, it can be very damaging. We often say things to ourselves that we would never dare say to anyone else. Our stress does not just come from external input, but our own emotional makeup can significantly add to it. In fact, studies have shown that our emotional makeup can actually cause so much stress in our bodies that it changes our physical make-up. Negative self-talk can cause changes in a number of health issues including:

- Diarrhea/constipation
- Muscle tension and pain
- Ulcers
- Headaches
- Insomnia
- Teeth grinding
- High blood pressure
- Cold hands/feet

All of these, in addition to our anxiety and depression, we may be suffering from. Chronic stress depletes the body of the essential chemicals that it needs to function properly. When we lack these essentials, we leave ourselves open for a number of illnesses. However, it is possible for us to change the way we respond to external and internal elements and develop a more healing environment that will give us better health.

It starts with changing how we talk to ourselves. CBT can teach you how to control your negative reactions to different things by teaching you different ways to interpret your personal experiences. So, rather than berating yourself for a simple mistake, you could remind yourself of all the times you did things well.

This helps to reduce feelings of guilt, shame, and anxiety that are eroding our mental and physical health. By changing these negative thought patterns into positive affirmations, you will find your mental state will gradually begin to turn into another direction.

Counter Negative and Angry Feelings

Anger is a unique negative feeling. Unlike other negative emotions like sadness, guilt, or disgust, anger can also have a positive side to it, so all anger is not bad. Sometimes referred to as the moral emotion, it often appears in cases where morals are brought into question. Values like justice, fairness, and respect are at its core.

There may be times, however, when you may not be able to make the connection between the root of your anger and its actual cause. You may, for example, feel you are angry at one person or another, but the cause may be from an entirely different source. This makes anger a sort of an enigma. We know that it usually precedes aggression, so it pays to address this issue as soon as possible. Negative anger can be treated with several different approaches to CBT. In fact, studies show that in as little as 8-12 weeks, there have been some very promising results. Techniques such as problem-solving, relaxation strategies, and enhanced communication skills have all been very effective in helping people to handle their anger better.

Dealing With Feelings of Guilt

Everyone has to deal with guilt at some point. No matter who we are, holding onto the wrongs we've done in the past is probably the most common reason why people don't move forward. Most of it comes from having to deal with the consequences of the poor decisions we've made, but the feeling that one act can stop us from moving on with our lives can be crippling.

No matter what has been done in the past, since it is not possible to do those things over in a different way, when negative results occur, the guilt that results can become like a life sentence that you impose upon yourself. Getting rid of that guilt on your own can sometimes feel impossible.

Through CBT, you can be guided through looking at that past event and learn how to see it through fresh eyes; eyes that can see the situation from different angles. One particular technique applied is a technique called The Blame Pie. This tool helps you to see just how much control you had over the negative situation and that, in most cases, you were not totally responsible. It looks at each individual involved in the incident and divides the blame according to their contribution to the event. As each person is assigned a percentage of the blame, it literally lifts much of the weight off your shoulders.

This type of treatment, over time, can help someone to get past the weight of the responsibility they had and finally find themselves worthy of moving on to a better life.

Counteract Hopelessness

Sometimes, depression can become so deep that a person feels completely hopeless. Because depression is the result of negative thinking, recovery from something as severe as hopelessness can be painfully slow. Even if you have positive thoughts on occasion, when you reach the point of hopelessness, you may not allow them to surface long enough to reap the effects.

Normally, in a depressed state of mind, positive emotions are often squelched with more powerful, negative thoughts like "I don't deserve to be happy," or "I know this won't last." In their minds, even when they feel good, they are unable to enjoy it because they are waiting for the ax to fall.

Through CBT, identifying the pattern of thoughts that lead to negative behavior can be changed. Once you're better able to identify these patterns, you can recognize them when they arise, freeing your mind for more positive thoughts. This can be done in several phases:

- Identifying the problem and mapping out solutions.
- Write down more positive statements that can counteract the negative thoughts.
- Actively search for new opportunities to apply positive thoughts.
- Learn to accept disappointment as a normal part of life.

Chapter 6: Steps to Dealing with Mood Disorders and Depression

Preventing mood disorders and depression is at the heart of Cognitive Behavioral Therapy. It has a multilevel approach that starts small and builds, helping patients to change their thought processes in everything they do. With each step, they celebrate small advances that will help them to gradually get closer and closer to their goals.

The first phase of the therapy involves identifying the problem they are facing, which is generally the result of automatic and intrusive thoughts. This is a crucial phase of the therapy as it defines a starting point for the treatment needed. Without a solid knowledge of the problem, it would be difficult to target the right strategies and techniques to deal with it directly.

The second phase is a period of goal setting where they can work on where they want to be within a set timeframe. After identifying the biggest issues you may be facing, it is much easier to map out a series of steps that will help you to focus on the changes you want to make.

In phase three, you will need to identify the challenges and obstacles you will be facing that will likely get in the way of your goals. No path will be easy as you will have to deal with a variety of setbacks along the way. You can decide to use them as a roadblock preventing your progress, as an obstacle that will require you to make a detour on your path to success, or as a stepping stone used to help you to get to where you really want to be.

Phase four deals with learning how to challenge your automatic and intrusive thoughts. In this phase, patients learn how to separate reality from the exaggerated thoughts they normally have. Once these negative thoughts have been identified, they can be challenged by a series of questions that forces them to use logic to come to a more reasonable conclusion. At this point, you will need to look for factual evidence that will contradict your embedded beliefs. Once those beliefs have been identified and successfully refuted, then you can learn how to use the evidence to refute those beliefs. Over time, those beliefs will diminish and be replaced by more realistic ones.

Phase five helps with identifying even more automatic assumptions and beliefs. Now that the automatic thoughts have been identified, this phase takes it another step forward and isolates the deeper core beliefs that all people have. These are the ones that are more absolute and less flexible. They most likely have been formed in childhood, and unless changed, will stay with you throughout your life, getting reinforced over the years with each experience you have.

This phase is the best stage to address these issues because many of us are reluctant to accept that these thoughts are wrong until they have addressed their own automatic thoughts first. Challenging these beliefs is often like challenging who you are as a person because they have been deeply embedded in your mind for years to come.

Phase Six focuses on changing the behavior. Up until this point, all of the exercises are dealing with the cognitive activity in the mind. These changes can now begin to extend outward involving participation in different activities and events giving special attention to those that will deliver a more positive mood.

For those dealing with depression, participating in these activities is crucial to lifting the mood. In this phase, the focus should rely heavily on avoidance issues, depression, anxiety, and phobias. Its goal is to boost self-confidence so you are more comfortable engaging in more meaningful activities.

Phase Seven deals with the problem-solving strategies you will need. It brings home the main point of Cognitive Behavioral Therapy; convincing your mind that everything is all right and that any problems that come up, you are capable of handling. While challenging your negative thoughts and beliefs will produce good results, the ability to master the skills taught in the problem-solving phase will be the most effective in helping to ease behavioral issues.

It is important to note that each of these seven steps can help you to move closer to your goal, it will not work well for everyone. Those who are suffering from severe depression, emotional trauma, and those mental illnesses that are considered to be extreme won't likely be enough. Most of the exercises and strategies you learn in CBT can be done and practiced on your own, but if you're dealing with extreme emotional instability, then you may need more guided therapy to help you get back on the right path.

The problem-solving phase involves several steps:

1. Identify the problem and the specific elements you need to address.

2. Brainstorm possible solutions.

3. Evaluate each solution to determine the pros and cons of each.

4. Choose an optimal solution and then choose a backup.

5. Create a plan of action.

6. Execute the plan.

7. Review results and make needed adjustments.

8. If the problem is not resolved, start again at step 1.

At this phase, you will also need to review your lifestyle habits. Modifications will likely be necessary as an aid to improve how your brain processes thoughts. You will be coached in matters of sleep, eating, physical activity, and meditation.

Most people will see impressive results after applying these methods over a period of several weeks. The most common approach is to take one or two weeks to address each step. After you have successfully met the challenges of one phase, then do not hesitate to move on to the next one. Hesitation could interrupt the momentum you've developed and could cause you to lose your steam.

Chapter 7: Multimodal Model

As you can see, there are multiple ways that CBT can help a person to counteract their negative tendencies. With their Multimodal strategy, the technique gets a little more in-depth with a closer look at the seven different aspects of the human personality.

With this model, therapy consists of using a variety of models at one time, all of which are able to be adapted to many situations because they are completely interactive. They also share the same core beliefs that an individual will use to build his thought processes on. They consist of a combination of several elements including:

- Their Biological and genetic makeup
- The Effects of their experiences
- The Sensations that exist between thoughts, feelings, and behavior
- The Imagery that occurs in their own minds, mental images both negative or positive
- Their Cognition, or the way they think whether positive or negative
- Their Interpersonal relationships with others
- Their Drugs/biology, or their physical experiences, or the use of substances

These are considered the core of this type of therapy. They can be easily identified with the acronym BASIC ID. Each of these modalities is used to help narrow down the areas in which the coaching sessions should focus on.

The beauty of the Multimodal Model is that it recognizes that different people will respond to different modalities. You may be capable of handling your problems on your own while others will rely heavily on the support of others, and others may prefer to deal with their problems through certain activities. All of these can be incorporated into the seven areas of the BASIC ID, so it is possible to focus on those strategies that will allow them to overcome the problems they face in a manner that they can feel more comfortable with.

The goal of MMD is to help the individual make the necessary changes so they can make the transition from their current state of mind to a more progressive and adaptable mental state. MMD is often used in highly complex cases of depression and performance anxiety. Those who benefit the most are experiencing negative behavior so extreme that their lives are paralyzed to the point that they are unable to move forward. Their careers and their families are already at risk or on the verge of falling apart. Some have a level of fear that has reached a climax that it has had a heavy impact on nearly all seven of the modalities, and an intervention is likely the only possible solution.

Its success can be attributed to the principle that by approaching several issues at the same time, an individual may find themselves dependent on substances or other crutches to deal with them. It could be a combination of health problems, emotional issues, and financial distress all at the same time. Each case severe enough to be addressed on their own but through MMD, it is possible to address all of these issues at the same time to facilitate a speedier recovery.

The Seven Modalities You Should Know About

Because most psychological problems tend to be multifaceted, it is necessary to start by understanding each of the seven modalities that are actually affecting you. When you consider that many of the intervention techniques that people naturally turn to are substances, it is helpful to give even more consideration to the last modality, drugs or substances, as they will hold many techniques that can reveal your inner self. Now, let's take a closer look at each one of them to see how they are used in CBT.

Behavior: This aspect of MMD takes a close look at everything a patient does. This includes his habits, gestures, actions, etc. Some of his behaviors will be healthy while others will not be. Unhealthy behaviors that require particular attention could be those that are destructive, immoral, childish, illegal, impulsive, controlling, or otherwise, inappropriate.

Most patients will seek therapy because their unwanted behaviors are causing problems in their lives. Practices such as overeating, excessive drinking, hoarding, rebellion, self-mutilation, etc. are the most common. The goal is to change the behavior itself through techniques like modeling, desensitization, and aversive conditioning. The trick is that nearly all unwanted behaviors are connected to the other modalities as well, so by including this in the MMD therapy sessions, there is a much higher chance of success and the prevention of a relapse.

Affect: This modality refers to the inner feelings and emotions you are experiencing. Throughout the process, you may feel a wide range of emotions, but the focus of the therapy is to address the feelings you don't like. So, while you may feel happy, sad, afraid, frustrated, and bored, the sessions will likely involve the negative emotions first. Many who are seeking therapy for other reasons will often find that these emotions are at the root of their problems, even if they don't realize it in the beginning. The negative emotions are usually the underlying trigger for a wide range of other feelings that are often buried deep inside.

Sensation: We have five senses – sight, hearing, touch, taste, and smell, which contribute to all of our physiological experiences. Negative sensations may be butterflies in the stomach, tense muscles, physical pain, rapid heartbeat, headaches, cold hands or feet, sweating, nausea, skin crawling, and shortness of breath. Some more extreme sensations could also be hallucinations and/or illusions.

Imagery: This modality consists of the mental images that are built up in a person's mind. It includes things they may fantasize about, their daydreams, and their own personal self-image. Common in those who suffer from anxiety disorders, their fears are part of their imagery. They also have an excessive worry about the future. Those who struggle with depression may have images that are extremely negative and distorted, far from the reality of things. Addressing one's imagery will help them to learn how to adjust the view of the world and give them a more realistic and accurate view.

Cognition: Focuses on a person's beliefs, attitudes, and judgments. When thoughts are negative like limiting or distorted beliefs about something, they can contribute significantly to depression, anxiety, or any number of other disorders. When a person believes they are not worthy of something, it can have an impact on their relationships, employment, and other areas of their life, which can be very damaging.

Interpersonal Relationships: This takes a closer look at how they interact with others, their social skills, how they relate to people, and what kind of support system they have or is missing in their lives. A close look at their relationships will reveal if they know how to develop and maintain a lasting relationship, feel connected to others, and if they have a good balance in mental health. Those getting over a breakup, needing to resolve conflict, or are antisocial will usually find that they are also lacking in the areas of effect and cognition as well.

Drugs, Health, Biology: This is a combination of several things. First, it includes a person's physical health. Whether he has been fighting off any serious illnesses, his overall physical condition, any physical limitations, his age, or chronic pain are just some of the health concerns they may have. It will also include biological factors like his genetics and brain chemistry and his need for medical treatment or medications. Finally, this modality looks at his lifestyle habits including diet and nutrition, activity level, sleep and eating habits, smoking, and drug and alcohol use. Close attention needs to be considered when dealing with substances. Few people realize just how much of an impact these habits can have on your mood or mental state. By evaluating this factor carefully, you may be surprised at what it can reveal.

There are two ways these seven modalities can be assessed. It is either with a one-on-one interview with a therapist or by filling out a Multimodal Life History Inventory questionnaire. Once an assessment is determined based on the BASIC ID, then a program of therapeutic techniques and strategies can be implemented starting with the modality that represents the biggest problem.

Chapter 8: Rational Emotive Behavioral Therapy

Rational Emotive Behavioral Therapy (REBT) places most of its importance on the thought processes of an individual. Like all other forms of CBT, it is based on the same premise that what we think triggers our feelings, which in turn, will trigger our behavior. The main idea behind REBT is to help those people who view their personal experiences in a negative way. However, they do not address the experience directly. Instead, the behavior is adjusted based on how they "perceived" the experience.

Through REBT, patients are taught how to challenge their own beliefs and replace them with a more accurate line of reasoning. It targets the inner beliefs, so they can deal with their experiences in a more realistic way. When successful, it can be very powerful, changing not just the way one thinks but also their perception of life in general.

REBT works because it recognizes that logic is not always an effective part of the human psyche. Even with logic, it is not always the best way to solve problems. Computers operate entirely on logic to perform their functions; they absorb data, analyze it, and using logic determine an acceptable output. The human brain, however, processes information differently, often without the use of logic. Therefore, some of their conclusions can be far removed from reality.

The focus of Rational Emotive Therapy is to teach people to think in a more logical manner. It works to break down the instinctual thinking process by using logical reasoning to interfere with the irrational assumptions they are making. Once these irrational thoughts have been replaced with more positive ones, it will start to filter down and change the inappropriate behavior to actions that are more positive.

No matter what negative thought a person may have, its roots are often deep within an unrealistic world. To combat this type of reasoning, REBT applies something called the ABCDE model of thinking.

A: Activating Event

An inciting event is identified which triggers the irrational belief. This could be any number of negative experiences including an argument with someone, a car accident, or the loss of a job. It is the trauma of this event that compels the mind to create a new irrational thought or belief.

B: The Belief

Once the belief has been created, the mind will automatically revisit it every time a similar negative event occurs. Each time the mind goes back to the belief, it is reinforced causing the person to get stuck in a spiral of negative behavior without ever really understanding why it happened in the first place.

C: Consequences

The belief will trigger the consequences of their irrational thoughts. Some consequences could be emotional as in the case of guilt or shame, while others could be behavioral as in overeating or some form of substance abuse. The underlying emotions for these behaviors could be depression, lack of self-confidence, or hostility.

D: Dispute

The dispute phase of the program is when you learn to challenge that belief system and see it as irrational. You begin to recognize it as the root of your problems. At this stage, you learn to argue with your subconscious mind and dispute your negative beliefs. You will be asked to come up with convincing proof that will successfully contradict your imprinted way of thinking.

E: Effect

You could also call this phase reinforcement. Once you have had your internal debate and successfully convinced yourself that your irrational belief is wrong, more positive behaviors will be much easier to come by. You'll feel a stronger self-esteem, you'll be bolder, or you'll just feel a lot better overall.

In most cases, REBT can be done without the aid of a therapist. This model can help anyone get to the root of their negative behavior and arm them with the tools they need to change. It motivates people to look deeply at how their thoughts are developed and how to apply rationale to their beliefs and replace negative thoughts with a more realistic view that will build them up rather than tearing them down.

Chapter 9: Dialectical Behavior Therapy

Dialectical Behavior Therapy is more like a positive form of psychotherapy that can be tailored to treat those with some form of borderline personality disorder.

More like a form of "talk therapy," DBT focuses on the psychosocial elements of treatment. Its theory is based on the premise that some people's behavior is so extreme that their reaction to romantic, familial, or social relationships is a manner that is far removed from the norm.

Those who fall into this category experience emotional extremes where their arousal levels can happen very quickly and their emotions are at such a peak that the reactions are at a very high level, taking them much longer to return to normal after an episode. They generally see things only in black or white with extreme emotional outbursts that can leave quite a bit of damage in their wake.

Because of this, they find themselves falling into one problem after another with no internal ability to manage their emotions. Therefore, they get no relief when their emotions spiral out of control. DBT is made up of three different elements:

Support-Oriented: Identifying their strengths and learning how to use them to help them feel better about themselves.

Cognitive-Based: Identifying the negative thought patterns and finding new ways to cope with triggering events so that their lives are more stable.

Collaborative: Rooting out underlying problems by working through assignments, active role-play, and learning self-soothing practices, so they can manage their own outbursts.

Through each of these sessions, the focus will be on two primary components:

1. Structured Individual Psychotherapy Sessions

Working with a therapist in a one-on-one session, patients will learn problem-solving behavior and openly discuss very specific issues that may have occurred in the past. These challenges could be anything from suicidal tendencies to self-mutilation. The more serious the issue, the more priority it should be given when working with the therapist. Minor issues may be assigned for the patient to deal with as homework.

During these sessions, you may also learn certain behaviors that can be applied to interrupt the negative habits. This adaptive behavior will have an emphasis on helping you to manage your emotions in the face of trauma much better. The objective here is to work towards a more socially acceptable behavior, so you can have better relationships with others.

2. Group Therapy Sessions

In group therapy, patients work their way through four different modules where they are taught practical skills they can rely on to help them cope with their negative patterns. Each skill has its own unique quality that can be used to minimize the negativity in their heads.

Mindfulness: A form of meditation where patients are helped to be more aware of their circumstances and the triggers that are causing their behaviors. Through mindfulness, they are taught how to observe, describe, and participate in their own thought processes as a neutral party. Then, with the use of special exercises, they learn how to recognize the triggers in their own mind, so they are better able to manage those reactions.

Interpersonal Effectiveness: Through practice sessions, they learn how to interact with others in a variety of scenarios. The process they go through is very similar to what is taught in many assertiveness training programs. Each session focuses on specific strategies to employ when getting to where they can ask for what they need, refuse things they don't, and specific coping mechanisms for when they are dealing with conflict.

Distress Tolerance: With distress tolerance, patients are given better ways to cope with disappointing and distressing circumstances. They learn how to accept things in a nonjudgmental manner. Rather than focusing on the negative, they will be better able to deal with real life situations that happen without their approval. Through this module, they are prepared with self-soothing techniques and other coping mechanisms that will reduce negative emotions and behavior.

Emotional Regulation: Learning how to regulate emotions is the key to managing negative behavior. Through the emotion regulation module, patients learn how to recognize the signs of oncoming intense emotions and use their coping strategies to manage them better. In this phase, they will be expected to address different emotional aspects including:

- Properly identifying and labeling their emotions when they arise.
- Recognizing the common obstacles that have been getting in the way of changing negative emotions.
- Lowering their susceptibility to negative emotions.
- Developing strategies to create more positive emotional experiences.
- Techniques that allow them to use their mindfulness to enhance positive emotions.
- They will also learn how to suppress their natural tendency long enough to choose another behavioral option opposite of what their instincts will do.
- They are given strategies they can utilize in distressful situations that will boost their tolerance levels.

With DBT, there are two main components, but because those that need CBT are most likely dealing with behavioral issues, they will focus most of their lessons in a group therapy environment, so they are able to practice their new skills and strategies they will need when interacting with others. DBT is still a relatively new program, but even so, it has already received recognition as a gold standard method of psychological treatment.

PART 2

Dialectical Behavior Therapy is a reinforcement that every life is worth living. This style of therapy is designed with your life in mind. These practices are not here to interrupt your well-being, but to reestablish what it is that you find wholesome and fulfilling about your life. This is a path that is ultimately led by yourself. This is your life at its fullest. The choices you make and new territories that you discover will be your own adventure, as set by your own path at your own pace. You have control over your own life, how exciting! By the end of this book, you will come to meet you as you in your most honest and most content being.

Chapter 1: Is This for You?

Before we start this adventure, we have to ask, who is this intended for? The short answer is that it is for everyone who wants to make a positive change in their lives. The key word there is "want". This is ultimately a choice. You must establish your own journey as the techniques exemplified in this book are just practices. There is no set number of meditation sessions that will unlock mindfulness. The practice of these techniques only increases the chances of your own self-discovery. Your willingness to find that goal is the only way these practices will be effective.

This may seem confusing or even overwhelming, but it should be celebrated! You have made a choice to better your life. You possess the bravery to examine yourself in your own state. You are already stronger for it. There is value in yourself and your life and you have already made the decision to discover yourself at your most honest, happiest state and to continue to not only endure but thrive in a world made by your own choices. The biggest step is the first one, and that step is already behind you. It is time to breathe a sigh of relief, to feel accomplished. The worst part of your journey is behind you.

Now that you have made the first step, where do you go? Obviously, the answer is your own choice. The practices in this book are merely there to help you along the way. This may or may not be a path that you have previously gone down, so use these techniques to guide you in your own journey. Look at this book as a toolkit. There is nothing in these pages that will assume a role of authority over you. That is the beauty of free will! You are free to explore at your own pace in your own order.

"Often, it's not about becoming a new person, but becoming the person you were meant to be, and already are, but don't know how to be."

— Heath L. Buckmaster, Box of Hair: A Fairy Tale

You have already made the most important step, and that step is the one that separates you from your furthest setbacks. There is already so much distance between where you were and where you are now. It is now possible to look back and accept yourself. Standing where you are now, it is possible to see your own worth. You are not your setbacks, and you are not your failures. In fact, you might be the most interesting person you know!

Chapter 2: Your Toolbox, DBT

The goal of Dialectical Behavior Therapy (DBT) is to separate you from behaviors that are harmful to yourself and others and replace them with meaningful habits. Now that you have taken your first step and have separated yourself from your setbacks, you can go even further and discover what it is that makes you truly happy on your own. Finding that you do not live to continue harmful behaviors but discovering and tailoring habits that will enhance the life that you are choosing to live will fill you with serenity and self-love, and it will be all the more meaningful because they will be your own interests and not the consequences of your setbacks. Honestly, how exciting is it to really discover the real you? Someone that you may have never met or may not have seen in a long time and neither has anyone else, a brand-new person who has been there all along.

The defined objectives of DBT is obviously a little more clinical. It includes Mindfulness, Distress Tolerance, Interpersonal Effectiveness, and Emotion Regulation. How does this relate to you, though? How do these skills fit into your new and exciting life? Remember that the goal of this kind of therapy is not to overtake your life, but to be there alongside it to help you discover what it is that makes you the real you.

Mindfulness is not a skill set, more so a state of being. Mindfulness is being aware of the present, in the present, and not to be overwhelmed by what is going on around you. It is an awesome way to be and reinforces who you are because only you have a mind like yours. Whenever you are using your senses to become directly aware of your present state of being, you are being mindful. Mindfulness is also exercised like a muscle. It is something that we all possess, but few regularly practice. Although that statement may not be true for long. There is a growing interest in meditation and a growing awareness of the importance of remaining mindful in every aspect of life from personal to even business. If you were to practice it, you will discover that the feeling of mindfulness becomes stronger the more you exercise that mental muscle. Focus and personal honesty will become stronger as you develop along this path. It is an exciting tool of self-discovery and one that will be explored upon later in this book.

Distress Tolerance is a measurement. It is your ability to accept distress that cannot be changed. Emotional pain is measured on a different scale altogether from physical pain, but it can be just as, or even more, damaging. The real skill here is learning how to find your own way around the distress and accept what you are unable to change. Practicing mindfulness will help you to separate yourself from distress factors but coming to terms with the reality of these stressful situations will no longer be a roadblock, but a defining challenge that will make you stronger and give you skills for future distress management.

"Grant me the serenity, to accept the things I cannot change; courage to change the things I can; and wisdom to know the difference."

Learning the difference between what you can and cannot control is paramount. Once you have accepted the reality of a stressful situation that you cannot control, you cease to try to change it but begin to find a path to live around or through it. Sometimes, the energy spent trying to change an unchangeable situation is more stressful than the original event! You owe it to yourself to not harm yourself. There are even times when the situation only seems to be so stressful because you have spent all of your energy and effort trying to change it instead of taking a step back and accepting it for what it is. You could even come to realize that the situation is more benign than how you have built it up to be inside your head. Sometimes, you can even find a way to turn it into a positive situation! You will never be able to do any of that if you are too busy stressing about the original situation, though.

Interpersonal Effectiveness will help you to build and maintain important relationships, including the one you have with yourself, as well as help you to define priorities and to arrange them in a sensible manner to live your new life the most effective. The clinical method is through the acronym DEAR MAN:

- **D**escribe the current situations
- **E**xpress your feelings and opinions
- **A**ssert yourself by asking for what you want, or by saying no
- **R**eward the person – let them know what they will get out of it
- **M**indful of objectives without distractions (attack the problem, not the person)
- **A**ppear effective and competent
- **N**egotiate alternative solutions

These are effective and healthy steps for conflict resolution and a great tool to have in mind to keep your communication on track and working towards an agreeable solution.

Respect is a trait valued by everybody in one way or another. Respect is earned and kept and can encourage stronger relationships with the important people in your life. Speaking in a respectful tone will lead you to your interpersonal goals in a way than getting agitated towards that person, situation, or even yourself. Self-respect is the true basis of interpersonal respect. Have you ever heard that you must learn to love yourself before you can love another? This is because you define for yourself, and exemplify to others, what respect means to you. How you treat yourself will set the standard for how others will feel that they can treat you. A person who dresses nice and speaks warmly with peers will garner more respect than a person who shows little care for how they want to be treated. Self-respect is important, and you deserve it! You are already stronger for having taken this journey and your story is one that no one else has. You are worthwhile, interesting, and unique. Taking good care of yourself will tell others that you are a person who warrants respect. Another acronym that is helpful about self-respect is FAST.

- **F**air to myself and others
- No **A**pologies for being alive
- **S**tick to values (do not do anything you will regret later)
- **T**ruthful without excuses or exaggeration

You have heard the Golden Rule; treat others as you would like to be treated yourself. Well, that rule works the other way as well! Treat yourself as you would treat others. You deserve the same respect that you would show to others, so do not count yourself out or make sacrifices that make you feel uncomfortable. Be fair to yourself!

If you find that you apologize unnecessarily, stop it! Sometimes, people will tell you that you apologize too much, which only make you feel uncomfortable. You do not have to apologize for anything that you are not truly sorry for. You occupy the same space as your peers and you deserve the same level of respect.

What are your values? Do you know? In your current stage of rebuilding and discovery, your values may change, or you may discover that you have been violating your own values for a long time. With a renewed respect for yourself and a bright new path ahead of you, you are most likely to find out what is truly important to you. Find your core values and remember that you deserve respect. You do not have to apologize for your values and you do not have to compromise your values. Make your identity known and remember that you are valid.

Once you know who you are, what you value, and the fact that you deserve and possess self-respect, honesty becomes easy. You do not have to fabricate yourself to fit in or hide any unsavory traits that you may think that you possess. Your peers will respect an honest you. Honesty to yourself and others is the pinnacle of freedom. You are who you are and who you are is a strong, healthy, and an interesting person! Half-truths and flat out lies do little more than create stress for everyone involved, including yourself. A person with self-respect does not need to create an identity that they do not own. Breathe and relax because you are you!

Chapter 3: Finding Yourself through Mindfulness

Discovering yourself is exciting! It's a journey that is enviable. We have already defined mindfulness, so the next step is to discover how it is practiced and define what your individual goals are. It is important to remember to constantly ask yourself what you want to find in this book. Your individual goals are the goals of this text. What practitioners of mindfulness usually find is greater fulfillment, a deeper understanding of their selves, positive behavioral changes, and more importantly, less suffering.

As you continue down this path, it is important to remember what your truest intentions are because doubts will surface. Mindfulness will need to be practiced and exercised like a muscle. Minds are messy, prone to wandering, prone to doubt, and everyone examines themselves much harsher than their peers would. In the last chapter, you discovered what your values are and who you are as a person. You discovered that self-respect is worth having. Now, it is time to reinforce what you know about yourself and what you want to explore.

Before we get to the actual practices, it is important to note that the path to mindfulness is not linear. It is a little different for everyone and the only outside guide is a collection of experiences from others. The true guide is yourself. Do not fret. Do not succumb to doubt because you may or may not discover a path differently or find a truth not listed in this book. No one can know you as well as you can. Instead of reveling in the doubt or confusion, be excited! You are the first to discover your exact path and you are the first to find your own unique solutions to your setbacks.

At the same time, you may discover that these goals are even connected! As you discover greater fulfillment, you may connect it to lesser suffering, and from there, you may find that you exhibit better behavior and more success in your relationships. Understand that practicing separateness from your suffering could lead to accepting validation from your own positive thoughts and energy.

The most obvious exercise for practicing mindfulness is meditation. It is important to note that meditation is not passive. It is not simply sitting and relaxing with your eyes closed. It is an active exploration of your mind while providing yourself with the least resistance to your own self-discovery. You may not just drop right into it during your first session. An unpracticed mind has never explored in that way. You may not know how to look inward as your senses and instincts are conditioned to look outward for stimulation.

First, you must separate yourself from your reactions. You must understand what your automatic reactions to a stimulus such as stress and joy are and be aware of yourself at the moment that you act automatically. You are not your feelings. You are not your reactions. Imagine you are on the side of a road watching traffic pass back and forth. Every car is a stimulus, feeling, or reaction. You are separate from them and you must merely make a mental note of it, and then let it pass. Be aware of their existence and acknowledge them, but do not react to them. Eventually, your mind will become more still.

Another example is to imagine your mind like a still pool of water. Every thought and stimulus is a pebble dropped in that pool. Those pebbles create concentric ripples that expand outward, and then even out. If you reach into the water to grab that pebble, you will only create a splash and larger ripples. Eventually, the pebbles will slow, and your quietest realizations and truths will surface. Do not fear! This is your truest self. This is exciting and another great achievement along with your journey to a more peaceful and successful you. After those truths have passed without judgment, your pool of water will fall even more still. You will experience true serenity and discover the most honest definition of a quiet mind. This is peace.

To practice meditation, you must first dedicate time and space to your session. You do not need a special pillow or certain music or any equipment whatsoever, just time and space to practice. Sit in a comfortable position that you will not stress to maintain and close your eyes. Next, just acknowledge the moment as it is. Observe it without judgment or interaction. Just simply be in the moment without exerting effort or energy towards it. Pay attention to the sensations of air passing through your nostrils or the presence of sound in your ears. Let the moment pass through you as you sit peacefully in it. The goal is not simply to be calm, it is to be aware of the moment as it is happening right now without interaction or judgment. The next step is not so much a step, but a reassurance. Judgments will rise. It is inevitable, especially when you are first practicing. Remain calm and remain practicing. Do not succumb to doubt or frustration. Simply make a note of it, and let it pass. This is an excellent practice for learning how to move on from frustration or feelings of grudge in your waking life. If your mind wanders too far off of your initial concentration, keep returning to the sensation of your breath. Focus on the gentle sensation of the in and out of your breathing. Simply be in your awareness.

Meditation is a proven method to reduce stress, increase clarity, and can even positively rearrange your brain chemistry! You will notice your brain will have less chatter in your normal life, and you will be less prone to anxiety. It is a great practice for finding a "third way" around a conflict. It can even open up your creativity and lower your heart rate and blood pressure. As I have mentioned before, it has even begun to appear in modern business practices. Some higher up CEOs have adopted this daily practice to increase their creativity and productivity and reduce their stress level in the fast-paced environment that is business. Everyone from athletes to political figures to your average working man benefits from this simple practice.

If you chose to partake in this particular practice, you are unlikely to regret it. The next chapter will focus on advanced meditation techniques for when you discover that you like this new calmer, more focused you!

Chapter 4: Taking Mindfulness to the Next Level with Advanced Meditation Techniques

If you have chosen to give meditation a try, then congratulations! You should feel proud of yourself for having the courage to try something new. You should feel reinforced in your feelings of solidity in your new and healthy life. You have made an actual effort and have taken real-life actions! This is another moment to look at just how far you have come. How has meditation affected your life already? Do you feel a renewed clarity? This chapter will show you advanced techniques that you can practice to further expand your meditation practices.

An easy form of meditation that you can incorporate in your daily life is called a Walking Meditation. Obviously, this can be done simply while walking, or any form of ambulation that you use to get around. It is an action that you do naturally and has been for years. You probably learned how to walk before you learned how to read! This kind of easy, almost automatic and steady movement is a perfect environment to study your meditation practices.

First, you must stand up straight. Keep your back straight as you practice this. It is important to find the posture that is comfortable and promotes easy steps and focus. Next, place your hands together just above your belly button with your thumbs curled in towards your palms. This position promotes a comfortable posture that brings your focus to your center. Your arms are not swaying, and you feel self-contained and comfortable. Now, let your gaze drop slightly. This will also allow you to focus while being aware of where you are walking to. Just like with normal sitting meditation, try not to get lost in outside stimuli, just simply make a note of them and continue on with your focus inwards. Now, you are ready to take your first step. In the last section, I mentioned that breathing could be used to bring your focus back to your center. In this exercise, you will use your steady footfalls to create a rhythmic cadence for you to keep your central focus on. Notice, without interacting, the sensation of the ground on your feet (or whatever mode of transportation that you would use to get around on your own). Notice as the ground rolls from the back to the front of your foot. Notice the gentle bounce of your body as you move along. Now, do the same with the next step and the next. Make sure to walk at a slightly slower pace than usual. It is not necessary to move ridiculously slow, just make sure that it is at a pace where you are able to focus on your gentle and rhythmic movements and still move along at a comfortable speed.

Benefits of this style of meditation are that it allows you to further exercise your focus outside of the room or environment that you have become comfortable meditating in. It allows you to start connecting that focus to your daily life as you practice maintaining that focus during the natural and unpredictable distractions that occur just in a day out. You will also begin to appreciate the seemingly mundane aspects of your day, bringing focus and renewed eyes to aspects of this wonderful life that may have gone unnoticed or underappreciated previously. A cloud moving in front of the sun might bring certain effects to your attention like the changing colors or temperature of this temporary state. You might find a renewed appreciation for the sun and life in general. A gentle breeze might remind you of how temporary forces in your life are. A passing conversation might show you how calm, focused, and centered you are feeling in the moment versus how frantic and anxious the average person is in their daily life. You will discover all of these things while keeping your focus centered. It is important to not react to any of their thoughts, just simply recognize the existence of these thoughts, and let them pass naturally on their own. Bringing your meditation practices from your sterile environment to the waking world is an excellent practice for learning how to maintain and call upon this state of focus when there are events in your life that may be exciting or stressful.

The next technique is quite the opposite. Instead of walking, this technique is most effective while laying down, but it can be done in a sitting position. It is called a Body Scan, and it is used to focus on your physical wellbeing. It gives the sensation of infusing your body with a healing breath.

First, you must sit or lie in a comfortable position. Do not pick a position or surface that will become uncomfortable or distracting during your meditation. Once you are in a good position, place your hands on your stomach in the same manner in which you did during the Walking Meditation, just above your belly button in a comfortable position that brings your focus to the center of your body in a full rest. Once you are in this position, you might find it easier to focus if you close your eyes. Now, take a few deep breaths. Take note of the moment as you are in it, just like you have practiced in the basic meditation technique. Then bring your attention to your body. Notice the sensation and pressure of the floor or chair on your back or legs. Keep taking deep breaths, but this time, notice the invigorating life that fills your body when you inhale deeply and then feel a deeper sense of relaxation on every exhale. Fall deeper and deeper into your focused state with each incremental breath. You may start to notice more minute sensations such as your pulse under your skin, or little hairs standing up on your arm as your body becomes more relaxed and focus.

Now, bring your focus to your stomach area. If your stomach is tense, let it loose. You might even notice that your entire body relaxes as you release the tension in your stomach. Shift your focus from your stomach to your hands just above that area. See if you can allow your hands to soften even more. Feel your body relax even another level. Now, bring your focus to your arms. Let the tension loose in your shoulders. Let the tension loose in your biceps and forearms. After that, it is time to bring your focus to your neck. Let the muscles in your neck relax. It is perfectly acceptable to let your body shift as your muscles become systematically more relaxed. It is almost bound to happen as you are achieving new levels of relaxation. How relaxed you are now will make your initial assessment when you first lied down seem so distant.

After you have relaxed your neck, then it is time to focus on your jaw. Let that tension go. In your waking life, the average person carries extra tension especially in their jaw, shoulders, and fists without even realizing it. You might perceive yourself as relaxed when in actuality; you are much tenser than what is comfortable. This is one of those realizations that you come across through meditation that is an invaluable lesson that you have taught yourself. After you have rolled your relaxing focus over your entire body, take a mental snapshot of your body as a whole. Notice your body in the same way that you notice passing thoughts in the basic meditation technique. You may realize that your body is yours, but it is not you. Your body is a vessel and a tool for who you really are. That separation is important when you practice meditation. It is what allows you to examine thoughts without attachment. Take one more deep breath and allow your eyes to open, feeling a new sense of invigoration and relaxation.

Congratulations! With these three meditation techniques; the basic meditation, the Body Scan, and the Walking Meditation, you are able to perceive and react to thoughts and stimuli within your mind, your body, and your world in a healthy way. There is nothing that you should not be able to process using these techniques. You now have the tools to tackle any hardships along your journey. On top of that, you now have a new perspective that is exciting to explore as you find new hobbies, relationships, and life choices. Now even simple tasks like breathing, walking, or even just existing can be healing and full of positive energy!

The next chapter will focus on processing negative thought patterns in a healthy way. Now that you have this new perspective and new tools, it should be easy to separate yourself from negative thoughts that may surface from your past or present life. Do not fear! You are ready. You are stronger than you have ever been, and you can tackle any setbacks you have experienced, or are currently experiencing. Take a moment to celebrate where you are versus where you have been!

Chapter 5: Using Your New Tools to Process Negative Emotions

Negative emotions will occur. It is the inevitability that comes with the endless possibilities of life. You cannot reasonably expect to live your entire life and never feel sad, hurt, angry, betrayed, embarrassed, or any other emotion that can be perceived as negative. In this chapter, we will review the skills that you have learned to more effectively process your emotions when an inevitably negative emotion occurs. Through Dialectical Behavior Therapy, Emotion Regulation breaks down into three goals.

1. Understand one's emotions
2. Reduce emotional vulnerability
3. Decrease emotional suffering

The first step begins with a simple truth, and that is emotions are not bad. Even negative emotions are not something to just be avoided. It is impossible, and unhealthy to attempt, to avoid every negative emotion that you will come across in your life. Attachment to negative feelings is what causes real suffering. You learned from the last two chapters how to separate yourself from thoughts and emotions. You simply must acknowledge the emotion and/or event, and then let it pass. It is important to acknowledge these emotions, though. Try to define your emotions clearly. Using phrases like "I feel bad" does not give a clear understanding of how you are feeling. Instead of "bad", expand on that. Pinpoint it by saying you feel frustrated, depressed, anxious, or angry. Understanding what and how you are feeling is integral to processing those feelings. It is also important to understand the difference between primary and secondary emotions.

Primary emotions are reactions to an outside stimulus, and secondary emotions are reactions to those primary emotions. For example, if you felt depressed later about being too angry at a friend, then anger would be the primary emotion while depression would be the secondary emotion. The secondary emotion is a judgment of the primary emotion. Learning how to acknowledge emotions without judgment is essential because secondary emotions are destructive. Also, learning how to process negative events without succumbing to negative emotions is very important. Maybe being angry at the friend was not the proper response when you could have used the DEAR MAN acronym in the second chapter of this book to properly resolve that event and those feelings in a way that would solve the issue and be beneficial to both you and your friend. Remember that emotions are not your identity. Emotions are there just to alert you to stimuli that are beneficial or problematic. How you process and express these emotions is entirely up to you.

Reducing emotional vulnerability will increase the stability of your emotions, simply put. In DBT, the methods for reducing emotional vulnerability is through action. It will teach you to create positive habits and experiences to balance out the negative feelings you might be feeling. An easy acronym to remember for this is PLEASE MASTER.

PL – represents taking care of your physical body and reducing or treating illness

E – eat a balanced diet

A – is for avoiding alcohol and drugs, which can only heighten or fabricate negative feelings

S – Sleep. It is important to get regular sleep

E – The last E is for exercise. Much like meditation, it will increase in benefits the more you practice.

MASTER – This one is the fun one. Master positive activities to increase your sense of well-being and accomplishment.

Your health affects your emotional state. This ties into the self-respect section that we talked about in the second chapter. You will feel much better physically and emotionally if you raise your standards of how you treat yourself. Getting regular sleep, exercise, and only treating your body and yourself to healthy food and activities will do absolute wonders for your confidence. This also includes avoiding alcohol and drugs. It is too easy to mask feelings with these substances, and as we have learned, that is not a healthy way to process those emotions. Avoiding emotions, especially with mind-altering substances, does not make those emotions go away. It is not a permanent solution, it only encourages you to chase that perceived temporary safety from those emotions while your body is developing an addiction to the actual substance. It is a trap and can only work to undo all of the work that you have already accomplished. Treat yourself better than that because not only do you have self-respect, but you deserve it.

Now, I am going to circle back to the PL portion of the PLEASE MASTER acronym. After you understand the steps necessary for taking care of your body, you will understand that it is important to monitor your body as a whole. This includes taking care of illnesses when they arrive. Illness is another inevitability of life. Much like emotions, it is important to process them in a healthy manner to avoid further damage. You deserve to live in a healthy body and you owe it to yourself to take care of yourself. Living in a healthy body will give you peace of mind. Knowing that at the end of the day, you are physically feeling healthy will put other situations in perspective and it will be one more positive that you can weigh against negative emotions when they occur. Along with exercise and meditation, you can choose to MASTER other positive activities in your life. Developing or rediscovering a hobby is exciting and can give new meaning and a new sense of accomplishment in your life!

After you have learned these skills, you are ready to learn how to decrease emotional suffering. In DBT, it is comprised of only two skills: Letting go and taking opposite action.

Letting go refers to what we have already learned, by using our mindfulness to process emotions in a healthy way by letting them pass without developing secondary emotions to attach to the primary emotions. Taking opposite action means engaging in actions that are in direct contrast to the negative feelings that you are experiencing. For example, instead of crying when a feeling of depression is acknowledged, try to stand straight, speak confidently, and react to the stimulus or event in a healthy way. This is not to ignore that emotion. It is an exercise to lessen the length and severity of the emotion. It is important to acknowledge emotions, but that does not mean that you have to be subordinate to them. You do not need to let emotions control how you think and act. It can also give you a new perspective on a situation that you may have reacted automatically too.

With these skills, coupled with the skills you have learned in the previous few chapters, you can process emotions internally in your mind, body, and everyday life and also express those emotions after you have processed them. Even more to add to that, you have developed a renewed sense of self-respect through self-care and new or rediscovered hobbies. You are now taking steps to replace negative habits and feelings with positive feelings and activities you enjoy and that are uniquely representative of you! You may start to feel that you are meeting the real you, a more positive and honest version of yourself, doing things that you enjoy.

Chapter 6: Defining Your Goals, Your Values, and Yourself

Now, instead of learning something new, it is time to reassess yourself after what you have already learned. Do you remember those goals and values that you defined for yourself at the beginning of this book? Well much like how we discovered new levels of relaxation during the Body Scan meditation, it is time to discover new levels of yourself. Maybe after you have practiced meditation and studied the different goals of DBT, your renewed sense of self and awareness can further sharpen your goals and expectations from your new life. It is even possible that you have already achieved and mastered some of your goals. If you have, then congratulations! It is time to reassess what is important to you and what you can get out of this book. If you have not achieved any of your goals yet, then do not worry! Hopefully, you have set expectations at a reasonable level and you are mindful of what you are able to achieve within yourself. It is good to have both long-term and short-term goals. It is important, even, to balance both so you are able to celebrate achievements along the path to a life-affirming goal that you may not have been able to achieve without taking that all important first step along this journey.

Each new skill you learn is a skill you would not have had if you would have maintained your negative feelings and habits. There are questions for you now that only you can answer. How is it that you feel? How do you feel in a general sense of wellbeing? How far along do you think you have traveled? You are most likely aware of your progress and it is good to celebrate along the way. These steps you are taking are not steps that any one person could take for you, no matter how influential or qualified. Just like how meditation and mindfulness is a study of you, the steps you have taken are entirely unique to you.

Having said all of that, it is important to allow positive feelings to be acknowledged and witnessed. Many have a hard time accepting themselves in their own achievements. Judgments upon oneself can absolutely be the harshest. It is easy for faults and negative feelings to seem large and overwhelming when you are standing so near to them. These negative feelings cause you to stress and can be impossible to simply ignore. This is why we learn to process those feelings and resolve them instead of trying in vain to ignore them. An unresolved negative feeling can trigger a survival response, which is why it is impossible to ignore said feelings. In this way, unresolved negative feelings make it near impossible to accept positive feelings about yourself.

Your body does not feel the need to react to positive feelings because it feels that the situation is resolved because it ended on a satisfying conclusion. Your body will tell you that your time and effort need to be spent resolving those negative feelings because they are triggering a survival response in you. Now that you have learned how to bring negative feelings to a positive and productive conclusion, it is now possible to accept your positive traits and individuality. It is even possible to meet yourself without those impossible stresses in your life. How exciting and life-affirming is that! How much better off are you now in relation to how you were before you took this journey?

Now, that you know your goals, and you know yourself, what are your values in your new life now? What have you learned that you could possibly maintain, or even teach others? Maybe you recognize the work that you have put in and are starting to recognize the results of hard work. Maybe you value patience and understanding because practicing meditation has taught you how to discover feelings that were always there, just buried. All this book can do is speculate and give examples to what you may be feeling. It is your unique journey that is your real teacher. You have taught yourself how to heal. You have taught yourself how to take the first step, and you have taught yourself how to recognize greatness within yourself.

Are there people in your life who would be proud of you for where you are now? If so, you should greet them and share your renewed sense of pride and clarity with them. It is reaffirming of your own sense of accomplishments to have it validated by those you love, those you admire, and those you respect. Sometimes, it can give a new perspective to emphasize with someone else and share a joyful feeling with them. You are no longer in a cave of your own misjudgments, both internal and external. You have established yourself out in the light. You can walk among the world with your head high instead of living in the past and inside of your own head. You see the world for how it actually is and not through the lens of prior transgressions or feelings of worthlessness. It is even possible to look back at how you used to exist and treat yourself and separate yourself enough from it that you can even brush it off. That is not you anymore. You are the real you now. You are the you that you were meant to be, a much happier and more honest you who recognizes real emotions instead of perceived injustices to yourself.

Chapter 7: Living in the Positive!

Now that you have created a positive atmosphere for your mind to exist in, you are probably feeling a new motivation and longing to explore the world in your new self. What do you do with all of this motivation? It is important to put this good energy to use as to not fall back into negative habits that your old self has come to reinforce. You are at a crucial step where you should give great importance to channeling this positive energy into positive habits.

Something that you can do for yourself is to continue to practice meditation and exercise. Your new positive life starts at your core. Your core being yourself. You have learned a renewed sense of self-respect and discovered some deep insight into yourself. Now, it is time to maintain it. You can continue to live positive as long as you take care of yourself. Imagine yourself in a fancy car. It can look nice on the outside, but if the engine is not kept in good condition, it will not function as intended. Every new positive action starts with a sense of wellbeing.

Other ways to maintain your emotional stability through practice is to find a creative outlet for your feelings. If you feel that you are the creative type of person, then you may already have some of these hobbies. You may even have hobbies that you have not visited in a long time. Picking up an old hobby can help you connect with who you were before you found yourself down a darker path. It can give you a sense that you are picking up where you left off and reassure you that this you is the real you. If you are not a creative person or have not found an interest in a hobby, then do not worry! Another way that you can strengthen your mental focus and reinforce this new positive you are to learn. Reading is a proven method to increase cognitive faculties and helps you to directly discover interesting perspectives that you may not have come to on your own independently. Maybe, you will even discover ways to learn about aspects of your life that you have put on hold. Projects and promises made to others and you can now be fulfilled because you are now breathing easy and have a new motivation for life.

Great! Now that you have a healthy and positive sense of wellbeing, you can further reinforce your new positive life by engaging in productive social activities. Before I get into examples of this, I want to further explore the benefits.

Giving back to your community outwardly shows that you want to be engaged with society. You recognize yourself as a part of a whole and you are devoid of an ego that alienates you from your peers. It is not a struggle for your individuality though; you have already explored and defined yourself to yourself. Now, it is time to show who you are to the world! A person who lives inside of their own negative thought patterns does not want to be a part of society. They will build their own mental walls to keep themselves from embarrassment, anger, shame, or any other negative thought patterns associated with social interaction. Maybe they feel that society owes them something. An overinflated ego is another trepidation to avoid. Now that you are free of all of these negative thought patterns, you can enjoy social interaction with a head held high and nothing to apologize for. Another key benefit that you may not have seen or realized before, is that doing something nice for others simply feels good. You are able to emphasize the happiness of others. Seeing a smile on another person's face that you have caused can feel so rewarding in ways that you have never felt before! Even for more selfish reasons, it feels good, as in the sense of being the hero of someone's day. It is a wholesome feeling. It is a feeling that is entirely guilt-free.

Some examples of positive social outings would be simple activities like volunteer work or attending or even participating in sports events. Maybe your place of employment has a softball league, or your colleagues enjoy disc golf. These are activities that directly give back to your community or peer group. These are higher levels of commitment, so if you are not ready for that quite yet, maybe you could try something a little less structured. Meeting trusted friends in a relaxed social environment could be a little bit more comfortable for you. Invite a friend, or a few friends, out for lunch or to a store of your common interests. This kind of setting makes for a good conversation that is not so personal if you are not ready for that. It is perfectly acceptable to take your time developing your social identity, as this step is very important. Meeting friends in this kind of setting can also help you learn more about your friends and even yourself! Maybe they have an interest that you did not even know that you had! Maybe you have a friend who is very interested in tabletop gaming, which might be an area of interest that you have never explored! Your friends and new interests will most likely lead you to new friends and even more positive and interesting activities. It is easy to get sucked into the positive life; all you have to do is take the first step!

Working within your comfortable level of commitment is essential, but it is also important to actually engage in these or similar activities. The goal of this section is to establish new positive habits to replace self-destructive habits. Just like how picking up and reading this book was a crucial first step, this is another crucial step. Do not fret though! This step is easier than what you would think. Most of the time, the fear associated with the activity is much worse than the actual activity, and you should know how to properly process negative thought patterns. All you have to do is breathe and take that step. Your friends, family, and colleagues will be more than happy to have you included.

It is important to establish positive relationships that engage in positive activities. It is also important to allow yourself to learn what positive social activities are. A common misconception, reinforced by advertisements and common television shows, is that all social activity takes place with alcohol. That is simply not true. In fact, the most productive and happiest people may rarely step foot in one of these establishments. As a side note, you may also be surprised at the money you save when you do not frequently visit these establishments, which brings me back to advertisements. That is why those media outlets pursue that lifestyle; it is purely to promote a lifestyle that will earn their company more money. In that respect, establish your own idea of happiness! Find out what it is that truly makes you happy! You are most likely to find that engaging with friends develops real bonds and promotes honest happiness. You are most likely to find that volunteer work, or even saying yes when someone asks for a favor, is more fulfilling than anything that you have experienced in your previous life.

You should feel proud of yourself for taking this step! Now that you are taking steps to not only better yourself, but to solidify and reinforce it with positive social activities, there is nothing that can get in your way on your path to being happier, more wholesome you! Once again, congratulations!

Chapter 8: How DBT Has Enhanced Your Life

Although this book has seemed to have an almost conversational flow, it has actually followed very closely to the five functions of DBT. As this book has mentioned before, the goal of this is not to assume an authoritative role over you, the reader. This book was designed to reinforce your own choices and merely give examples of positive living for those who may be unaware or fearful of how to live as such. Having said that, it is now time to relate what we have learned to the five functions of Dialectical Behavior Therapy. Before we do that, let's define what those five functions are.

- Enhance client's capabilities
- Improve the client's motivation
- Assure generalization to the client's natural environment
- Structure the environment
- Enhance the therapist's capabilities and support their motivation

The clinical way to go about enhancing your capabilities is to reinforce the skills of DBT. We have used many skills directly from the actual standard of DBT, such as the acronyms DEAR MAN, FAST, and PLEASE MASTER. These are acronyms that you would become familiar with if you were to attend a regular DBT session. We have also discussed important skills like practicing mindfulness and emotional regulation. These are also skills that you are most likely to encounter in an actual DBT session. This book has taken those lessons and broken them down for you to study, practice, and make into your own at your own pace by your own choices.

Using these skills in your own life will only work to enhance the quality of your life and introduce you to lifestyles that mirror your interests, even those that you may not be aware that you have. This is an exciting time to be alive, and an exciting time for you!

The next function of DBT is the enhancement of the client's motivation. This book was designed to keep you motivated throughout, but it is not what was written or the speed in which you read it. The real motivation comes from you. You have rewarded yourself for picking up this book and sticking to it all the way to the end. By now, you deserve to have developed a sense of pride in making these positive changes in your life! There is no outside force that can motivate you to the extent that you can motivate yourself.

This book was written to be a companion to your own life. You are free to read or not read, follow or not follow, at your own pace. The fact that you have made it this far is something to be celebrated. It shows that you are honest in your desire to rid yourself of negative thought processes and self-destructive habits. There is not a single person or entity that is able to instill that level of motivation inside of you. You have shown that you are committed; not to this book or these processes, but you are committed to yourself. You have already taken better care of yourself than previously you might have thought possible. It is not only acceptable but appropriate to celebrate yourself at this time. This is a real achievement that you have accomplished, and one that many people take multiple tries to achieve. Some may not ever get to the level of clarity and health that you have already achieved for yourself. Once again, Congratulations!

The third function of DBT may seem confusing at first. It is to assure generalization to the client's natural environment. What that means is that this treatment, and this book, is designed to be a companion piece to live alongside without overtaking your life. This is not a program designed to put your life on hold. The effectiveness of this is that it promotes ease of transition into your new lifestyle while giving examples that are digestible by you because they relate to you, just as you are. It is easy to take this book with you and read it at your own pace or use the skills you have learned through a DBT session or in this book while you go about living your day-to-day life. There is no commitment besides the commitment that you have made to yourself and are comfortable with.

In an actual DBT session, they would address this function within the moment coaching. You would have access to a 24/7 phone number that you would be encouraged to call if you are having a hard time with applying the lessons to your life you would have learned during a session. This is an excellent tool, and if you were to attend a DBT session, I would strongly encourage you to feel free to use it. These coaches are not there to judge your choices. They understand the material and are also encouraged to process emotions without judgment. This is purely for the benefit of you! It is also encouraging to have outside motivation when your motivation might be hitting a low point. There is nothing to worry about though, just like the inevitability of sickness or negative thoughts, you cannot fault yourself for when your motivation is feeling weaker at the moment. Just relax, call that number, and celebrate yourself for making the positive choice at that moment when you may not have previously.

We are almost through the list here, I hope that you feel encouraged to continue. The fourth function of DBT is to structure the client's environment. This one can seem almost scary because you have gone this far along your own choices. There is nothing to fear though because this step is not designed to take away your choices, merely to help you and provide tools for you to make positive choices when you may not have previously. How a DBT session would go about doing that would be to assign you a case manager. This is someone who is dedicated to your case and is working with you to ensure your success.

An important aspect of this function is the thought process when accepting it. It is not there to control your lifestyle. When a client has made poor life choices and made a habit out of them, then they might not be aware of or be comfortable with lifestyle choices that are more positive and sustainable. You have already decided to live a positive life, now it is time to learn how. That is the purpose of this function. In DBT, there is a strong focus on replacing negative habits with more positive habits. This is because pure motivation has to be outwardly expressed and used for it to continue. Imagine your positive motivation as a match. You can light the match, and it will burn for a while. It is hot, it is bright. It has the potential to continue on, but it can only continue on if fuel is introduced to the match. Imagine this function of DBT as a pile of wood arranged for you in a fire pit, ready to be lit by your motivational match. Once you apply the match to the wood in the fire pit, then the fire burns much longer in a safe environment. Your motivation must be applied to a positive atmosphere to continue on. Your case manager or other individuals in your DBT session use this function to safely provide you with those structured, positive environments. Go forth and do well for yourself and others!

Have you made it this far? I hope that you have because this is now the final function of Dialectical Behavior Therapy. That function is to enhance the therapist's capabilities and support their motivation. DBT therapists work in a team to more effectively enhance the lives and understanding of their clients. This is important for the team as well as the client. A typical DBT team meeting may start with a mindfulness exercise, reading of the previous minutes, and then discuss strategies to further their treatment. It is important for you to be engaging and helpful along with your therapist as this whole treatment only works with your commitment. An example of this would be to imagine you and your therapist on a rowboat. Your therapist will not be able to motivate you to continue to row if they are not participating in the work. Your therapist can also not row by themselves if you are not helping. This whole style of treatment is designed to be a cooperative endeavor. You should feel excited and encouraged to participate. The end result will be a happier, more positive you!

This chapter is here to serve the purpose of relating what you have learned to the structured skills that are discussed in an actual Dialectical Behavior Therapy session. It is strongly encouraged that you attend these sessions and take what you have learned in this book with you to those sessions. There is nothing that you should not be able to achieve in this aspect of your life between this book, those sessions, and your own motivation! You have a threefold angle of attack on your negative habits that you wish to eradicate from your life. Finding a DBT session is easy, as it is a growing style of treatment. Everyone involved wishes only the best of success for you! Continue on with your own choices and feel proud of how far you have come!

CONCLUSION

Dealing with psychological disorders of any kind can be difficult. Anyone that is struggling to live with them can attest to the challenges they face. Hopefully, as you have read through these pages, you've learned some practical tips on helping yourself to heal. In fact, you may have learned quite a bit about yourself in the process.

CBT is one of the most progressive forms of psychotherapy being used today. Unlike the Freudian form of therapy that has been used for centuries, CBT's primary focus is on identifying the thoughts that trigger the negative emotions and behaviors and employing strategies to change them.

While the techniques and strategies discussed here are usually done with the help of a therapist, as you can see, many of them have very basic concepts that can be adapted to fit a wide range of situations. By learning to go through these exercises on your own, you can effectively heal yourself through your own mental powers.

The concept behind CBT is very simple. If we change the way we think, we can change the way we feel. If we change the way we feel, we can change the way we behave. Our brains are highly complex machines. If we find ourselves in a situation that we have a difficult time understanding, it will begin to create a scenario in your mind that will fit closely to that situation. This is a natural coping mechanism to protect us from the pain of disappointment. So, if we are dealing with trauma, your brain will figure out a way to get through it. The problem, however, is that when your perceptions of things are wrong, the automatic and intrusive thoughts that are generated begin to trigger negative emotions and behaviors that are out of sync with the rest of the world.

As a result, we have problems that are difficult to overcome, which can compound into even larger problems creating a snowball effect that may seem almost impossible to overcome. By employing these strategies to redirect our minds, it will be like learning to think all over again; learning how to be more positive, to live in the real world and not in our imaginations, and function in a way that we can grow from.

Through the pages of this book, we have discussed the basics of CBT and how to begin to heal yourself. Unfortunately, we have only just begun to scratch the surface of this topic and we encourage you to continue your quest to find more information about it.

Now that you know what to do, the principles behind the therapy, and the steps to take, you can be on your way to better mental health. We close out this book with one admonition. The fundamentals of CBT are very simple; implementing many of the strategies here does not have to be difficult. For most people, it can be a method of self-healing. However, there are some who will struggle with even the most basic of these ideas; their mental state is much more complex.

For those people, we encourage you to take a step further. It will be worth it to find a reliable therapist to aid you in your efforts to get better. If you find that using this information is not getting you to where you want to be, then don't hesitate to reach out for more help. It is out there, and there are many professionals that are more than willing to help you get over the hurdles.

There is a reason why you chose this book. If you are feeling alone or lost in a world of billions, or you just can't seem to get out of your rut, applying these basic lessons can be very instrumental. The sooner you start, the sooner you'll begin to start feeling better. So, what are you waiting for! It's time to reclaim your life and get back to living the way you were meant to live.

Anxiety is something that every single person must face, even when they least expect it. You can either continue to give it the power to wreak havoc on your everyday life, or you can choose to do your best to lessen or even eliminate it completely with the techniques and strategies you have just learned from the previous chapters.

My hope is that by educating even just a few people on the power of taking back control of your life with self-awareness, positive thinking, and re-engineering the mind, that our world can become a better place, one healed mind at a time.

You have truly made a huge impact on yourself, and quite possibly the biggest positive impact that you have made thus far in this journey of life. As I have said in the introduction, this was a journey made by your own choices. The secret about adventures is that they are never really over, and it is the journey, not the destination, that is the most exciting.

By the end of this book, you should have developed positive skills to replace your negative habits. Then you have applied those skills to your life in the form of practicing mindfulness, meditation, and Emotional Regulation. Then, you took it one step further and examined your life as it is now and developed the motivation to pursue social activities that reinforce your new positive life. This is a major victory for yourself and you should feel so proud of where you were, where you are now, and where you are most likely headed!

Like I have mentioned, it is strongly recommended to continue on in a DBT session. They are designed to work alongside you so that you do not have to put the progress that you have made on hold. The sessions usually take about a year to complete, but it will not be interfering with your life. This is your life at your own pace. You are most likely breathing easier, living better, and most importantly, feeling much more positive about you, your lifestyle, and your choices. Your family, peers, and colleagues are most likely excited to have the real you back, and it is now possible to continue on living without the weight of negative thought processes holding you back.

"With the new day comes new strength and new thoughts."
- Eleanor Roosevelt

I hope that this book has been a successful companion to you in your journey to discover the real you. Remember that this is not someone else's life that you are living. These are not processes designed to make you feel a certain, exact conclusion, but to merely enhance what it is about you that makes you the real you. Everyone has a positive side to them, everyone has a reason to celebrate themselves. Now, you are able to do so without the added pressure of negative thought processes, negative emotions, and negative habits. Go out and share your new self with the world. You are ready to become the best version of yourself!